MW01156657

Geothermal HVAC

About the Authors

Jay Egg (Port Richey, Florida) started Egg Systems in 1990 to provide energy-efficient geothermal air conditioning systems to the Florida market. Since then, Egg Systems has successfully installed thousands of geothermal HVAC systems for residential and commercial customers, in the Tampa, Orlando, and Atlanta regions—and beyond. Jay and his team are currently working with many exciting engineering projects and developments internationally. Jay grew up in California and served in the U.S. Navy as a Nuclear Field Electrician. Jay trained with Dr. James Bose of Oklahoma State University, whom many consider the father of the modern geothermal HVAC movement in America. Jay appears frequently on TV and in print, and regularly authors papers and speaks on the topic.

Brian Clark Howard (New York, New York) is an environmental journalist with a decade of experience in websites, magazines, books and other media. He serves as a Web Editor at The Daily Green (www.thedailygreen.com), which is part of Hearst Digital Media and is one of the world's largest and most trusted sources for consumer information on living a more environmentally friendly life. Brian was previously Managing Editor of *E/The Environmental Magazine*, the oldest, largest independent environmental magazine in the U.S. He has written for *Plenty, The Green Guide, Miller-McCune Magazine, Popular Mechanics* online, *Men's Health, Mother Nature Network, Oceana, AlterNet, Connecticut Magazine* and elsewhere. He wrote the chapter on saving energy for the 2009 book *Whole Green Catalog* and the chapter on green power and green lighting for the 2005 book *Green Living*, which he also coedited. He coauthored *Green Lighting*, the recently published McGraw-Hill book. Brian has bachelor's degrees in biology and geology and a master's in journalism from Columbia University. He was a finalist for the Reuters/IUCN Environmental Media Awards. Brian has appeared on numerous radio and TV programs and blogs for AOL's Asylum.com and as the URTH Guy at The Daily Green.

Geothermal HVAC

Green Heating and Cooling

Jay Egg
Brian Clark Howard

New York Chicago San Francisco
Lisbon London Madrid Mexico City
Milan New Delhi San Juan
Seoul Singapore Sydney Toronto

The McGraw-Hill Companies

Library of Congress Cataloging-in-Publication Data

Egg, Jay.
Geothermal HVAC : green heating and cooling / Jay Egg, Brian Clark Howard.
p. cm.
Includes index.
ISBN 978-0-07-174610-6 (alk. paper)
1. Ground source heat pump systems. I. Howard, Brian Clark. II. Title.
TH7417.5.E38 2011
697'.04—dc22 2010028695

McGraw-Hill books are available at special quantity discounts to use as premiums and sales promotions, or for use in corporate training programs. To contact a representative please e-mail us at bulksales@mcgraw-hill.com.

Geothermal HVAC: Green Heating and Cooling

1 2 3 4 5 6 7 8 9 0 QFR/QFR 1 9 8 7 6 5 4 3 2 1 0

ISBN: 978-0-07-174610-6
MHID: 0-07-174610-2

The pages within this book were printed on acid-free paper.

Sponsoring Editor
Judy Bass

Acquisitions Coordinator
Michael Mulcahy

Editorial Supervisor
David E. Fogarty

Project Manager
Aloysius Raj, Newgen Imaging Systems Pvt. Ltd.

Copy Editor
Neil Otte

Proofreader
Anula Lydia, Newgen Imaging Systems Pvt. Ltd.

Indexer
Ken Hassman

Production Supervisor
Pamela A. Pelton

Composition
Newgen Imaging Systems Pvt. Ltd.

Art Director, Cover
Jeff Weeks

Contents

Acknowledgments

Jay: Thank you to Seth Leitman for finding me through the Web and asking if I would be interested in writing a book, and then recommending Brian Clark Howard to cowrite. Brian's wealth of knowledge on all subjects concerning the environment has been irreplaceable, as has his patience and encouragement.

I would like to thank Judy Bass for seeing far beyond my vision for this book. I certainly never expected to be a writer, and her encouragement was all I needed to start.

Thank you to Tom Cavanaugh, currently serving as a mission president for the Church in Colombia. He counseled with me on the difficulties that I would and did endure in the process of writing this book. Thanks to the folks at ClimateMaster, Dan Ellis and especially John Bailey, who spent a good deal of one-on-one time with me to answer some tough questions.

My entire staff at Egg Commercial have been so very patient with me. A big thank you to Christina Brewer, our chief executive; Kristin Sagert, our executive office administrator; Jason Hodges, our Webmaster and marketing genius; and Sarah Cheney, our illustrator.

No words can express my gratitude to my family. They have seen little of me these past five or six months. Thank you to Kevin and Katie; Jordan, my 16-year-old beauty queen; Taylor, my strong 14-year-old son who still can't beat me in wresting; Hannah, my 11-year-old beautiful little pixie, and Theron, my 10-year-old all-American boy and fourth-generation namesake. My heart goes out to the lovely Mrs. Egg... I would have quit long ago if not for her absolute, unending love and faith in me. Thank you baby doll. This has been an entirely encompassing endeavor of love.

Brian: First of all, thank you to Jay Egg, who has been exceedingly gracious in sharing his wealth of experience and in-depth technical knowledge. Working with Jay has been a great pleasure, and I am deeply inspired by his passion, commitment, and Herculean work ethic. Without Jay, this book would have been impossible. I would also like to thank Judy Bass and the McGraw-Hill team for their professionalism and unwavering support. Thank you to Seth Leitman, who honored me with the invite to help with this book and serves as a valued group editor.

I would like to thank my mentors, Doug Moss and Jim Motavalli, who taught me so much about the possibilities of going green and life, at *E/The Environmental Magazine.* As Jim told me when I started as an intern ten years ago, it really is possible to change the world. I also need to thank my brilliant colleagues at The Daily Green, Dan Shapley and Gloria Dawson, who teach me and inspire me every day. Thank you to Remy Chevalier (remyc.com), who has taken many hours to explain complex topics to me and who has steered me to many invaluable sources. I'd also like to thank all my friends in the green blogosphere and throughout the green movement. There are too many names to list here, which perhaps is a testament to how collaborative, supportive, and creative this space is. Every day I am honored to be a part of it.

I'd also like to thank my family, who taught me to respect and appreciate the natural world and to strive to leave everything better than how I found it. That's something I know Jay can appreciate. Thanks to my parents, Allan and Diana, and my sisters, Amy and Lisa. Thank you also to my wonderful, beautiful girlfriend, Gloria, who supports and challenges me.

Introduction

I have been passionate about geothermal air-conditioning for 20 years because it is renewable, sustainable, and comfortable. And now with new federal incentives in the United States, I am glad to say that I can add a fourth watchword: *doable*. Even so, my favorite word of the four is *sustainable*. You may think it's because geothermal technologies can reduce our reliance on dirty, scarce fossil fuels and lessen our impact on the environment. That's certainly true, but that's not the first thing I think of.

You see, I grew up as the oldest of nine children on a nine-acre ranch in the high desert of California. My father is a great man, a high school English teacher, who provided well for our family on $18,000 per year during the 1970s, my formative years. Our electrical budget was $50. Our food budget was $200. We had a shared party line for the phone and no cable TV. We could get one or two network affiliates by antenna if the atmospheric conditions were right. By our very nature, we were green before it was cool.

If we wanted hot water, we had to make sure the black garden hose was spread out just right on the roof and turned on to fill the hot water tank. We never had a dryer. A warm fire in the winter and a swamp (evaporative) cooler in the summer in the common room were the total extent of climate control. Most of our fruit and vegetables came from the garden. We had livestock for milk and food, including chickens for eggs and the occasional Sunday dinner. We were able to sustain our lives on what we had. To this day, that's what "sustainable" means to me.

I now live in a 3000-square-foot home on a tad over two acres in Pasco County, Florida, not far from Tampa. My beautiful wife and I have four children, a garden, goats, chickens, pigs, dogs, cats, and a solid desire to maintain family values and a sustainable lifestyle. We live a very comfortable life, with every luxury I could ever have imagined, but sustainable. A wise leader in the Church once provided counsel that I can't forget: When you buy an item, luxury or not, the retail cost is only a fraction of the real cost.

Picture a boat, an ATV, or an RV. The real costs are quantified in the time and resources it takes to fuel it, maintain it, repair it, clean it, store it, insure it, license it, and then advertise and finally sell it for a loss—or dispose of it properly. It turns out that, in most cases, these are not very sustainable items.

The same goes for many landscaping items, such as nonnative sod and shrubbery. I have a beautiful, natural Florida landscape with native grasses, pine trees, and scrub oak. My Bahia grass, which you often see along freeways here, requires only monthly mowing. Imported grasses require constant watering and mowing and influxes of fertilizers and pesticides, although they still get brown spots from dryness and cinch bugs. Sure, I have been tempted many a time to install a sprinkler system and a "perfect" manicured lawn, or some exotic flowering plants like the neighbors have. Then the idea of sustainability comes to mind. If I get such items, nice as they may seem, there's an ongoing cost to maintain them, water them, fertilize them, protect them from pests, and so on.

I could afford all of the above items. But I choose not to have a boat, an RV, or a carpet of green grass, or any of the costs that go along with them. What I have is a house with a solar water heater, geothermal air-conditioning, fluorescent and LED lighting, beautiful garden, and a few farm animals. I have five bedrooms and five bathrooms, a 15,000-gallon pool, and a 1500-foot lanai. My energy bill averages about $250 per month. I have no water, sewer, cable, or phone bill. I call that *sustainable*. The money I save goes to saving for retirement, schooling, children, charity, and a good dose of spontaneous family fun. Things like eating out, vacations, and visiting theme parks and museums. These are things that bring our family closer together, truly enrich our lives, and sustain our economy.

Sustainability is what this book is really about. I have one employee for whom we are sizing a geothermal system at the time of this writing. He has a 2000-square-foot home with an average monthly electric bill of $450, on top of water, sewer, and trash. That is not sustainable for him. It is so expensive that he may never get ahead. With utility bills rising beyond inflation, he may be facing $600 to $800 bills in the next few years. Many others are in this situation.

Typical air-conditioning (AC) equipment, which has a seasonal energy efficiency ratio (SEER) of 10, is sucking the life right out of our fellow consumers. The cost of air-conditioning systems has doubled over the past five years. This is partly a result of rising material costs, but is primarily due to government-mandated efficiency requirements for 13-SEER equipment. A 30% increase in efficiency makes a big difference, but it's not enough. The federal stimulus package has allowed for $1500 tax credits for air-source (nongeothermal) residential systems that meet certain criteria. That stops at the end of 2010. After that, the only air-conditioning and heating product left on the stimulus radar is the little-known geothermal system, the subject of this book. The tax credit for geothermal heating, ventilation, and air-conditioning (HVAC) will be in force until 2016, offering an unprecedented 30% of costs, with no cap on the amount of credit you receive. There are even more lucrative incentives for businesses to install the technology, as we'll soon demonstrate.

True, geothermal systems tend to be a bit more expensive upfront today, versus conventional HVAC alternatives. But once the infrastructure is in place, it is very easy, and cheap, to maintain, and it pays for itself over a few years. The air conditioner component shouldn't need to be replaced for 20 to 30 years, and then you just replace the central unit, not the whole piping system. That's easy and affordable, since the drilling aspect is responsible for much of the cost. Geothermal HVAC systems are exceptionally quiet and reliable, and they tend to produce especially even heat or coolness.

Judging by recent developments, there is a good chance that geothermal heating and cooling systems will become a required standard in at least some sectors of the building industry before the end of the decade, and possibly by 2016. Unlike solar or wind power, geothermal works 24-7. It is highly effective at reducing the peak demand that stresses our power grid, thereby reducing the need for new power plants. Geothermal is certainly an effective way to reduce our greenhouse gas emissions.

Whether you're a contractor or a consumer, you should be able to pick up this book and learn what to expect from geothermal HVAC, the common pitfalls to watch out for, what to say to whom, and how you can benefit from the quality and long-term savings of the industry. We organized it so you can read it cover to cover, browse through the pictures and captions, or use the index and table of contents to get right to what you are looking for. We hope you'll discover why we believe geothermal is the way of the future, and one of the most promising and exciting technologies available today.

—Jay Egg

One note about how this book is written: All sections presented in first person are by Jay Egg, unless otherwise indicated. Parts that use the plural "we" refer to both Jay Egg and Brian Clark Howard.

CHAPTER 1

Introduction to Geothermal Technologies

It may be hard for an egg to turn into a bird: it would be a jolly sight harder for it to learn to fly while remaining an egg. We are like eggs at present. And you cannot go on indefinitely being just an ordinary, decent egg. We must be hatched or go bad.

—C.S. Lewis

I grew up on historic Route 66 in the Mojave Desert of California. If you've ever seen the Disney movie *Cars,* that was basically my stomping grounds. During the long, hot summers, we would often retreat to the comfort of our underground forts. These one-room caverns were cool and comfortable, and we would adorn them with furniture and even lights. An extension cord would be run from the closest power source. A sofa would be procured from our ranch-style home. And we would even have running water thanks to the nearest garden hose. Often the temperature was well over a hundred degrees in the summertime, so hot that we could barely drink fast enough to keep ourselves hydrated while in the sun. However, we could actually nap in these awesome underground retreats during the heat of the day. The same thing happened in the winter. With the temperature dipping into the teens, the underground forts were a cozy meeting place.

We got the idea while digging a septic leach line by hand with Dad in the backyard one summer. When lunchtime came, and mom brought out the sandwiches and lemonade, we would settle down in the cool earth at the bottom of the 8-foot deep trench. Dad asked why we didn't come up, and we explained that it was just too hot up there. That summer, we commissioned our first underground fortress, completely dug by hand. It was large enough to park a full-sized truck inside. The roofing was corrugated steel, which we then covered with a foot or two of the earth we had excavated from the hole.

Years later, I visited caves in Kentucky while on vacation. This was pre-1990, prior to my first experience with geothermal air-conditioning. I remember thinking, as I entered a cave and left the blazing, 96-degree heat of the surface, that it seemed a shame that we didn't build homes underground. The temperature was truly wonderful, and I thought at the time that I would like to explore the possibility of subterranean climate control as an energy-saving measure. Though I was determined to pursue my theories further, life got in the way, and I didn't think about it much until several years later.

"Discovering" Geothermal Cooling

I now live in Florida, and although the summer temperature doesn't often get above 92°F, the humidity is often around 100%. On Labor Day weekend in 1989, the compressor on my 3-ton Trane air conditioner failed. Since it was a Saturday, I was determined to wait until Monday, when the HVAC wholesale house was open, and I would be able to purchase a new part. Just a few hours later, it became apparent that there was no way the home was inhabitable without an operating air conditioner. I dialed the on-call number for the parts house and agreed to pay the $25 fee for the attendant to meet me there and open the store. I returned and installed the compressor in about an hour.

At that point, I noticed that the heat coming out of the condenser was intense, and that the liquid refrigerant line going into the house was almost hot to the touch. This should optimally be just a few degrees above ambient outdoor temperature. The condenser was drawing 18 amperes (A), which was right at its full load capacity.

As I considered the severe conditions that likely caused the failure of the air conditioner in the first place, my lawn sprinkling system cycled on in the backyard. I looked at the seemingly endless supply of water coming from my shallow well pump and got an idea. I checked the temperature of the water: It was 72°F. So I went over to my service truck and found a water-to-refrigerant exchanger used for a three-horse-power icemaker (Fig. 1-1). I installed it in place of my outside, air-cooled condenser coil, using water from the well. This took about two hours and some creative engineering, but it went fairly well.

I turned on the system and found the liquid refrigerant line was very cool to the touch (see Fig. 1-2). Even better, the power draw dropped down to 9 A from 18. The air coming out of the grilles was almost 10 degrees cooler. And the humidity dropped significantly, from about 70% relative humidity (RH) to about 55%. There it stayed for the remainder of the summer.

At this point, I concluded that I may have become the inventor of groundwater-cooled air-conditioning (I wish!). I went to work on this idea, and after a short time I had a good design and was ready to find a manufacturer and a patent attorney. That was when a wise old contractor

Figure 1-1 Water-to-refrigerant exchangers like this one are the heart of many geothermal air-conditioning systems. (*Sarah Cheney/Egg Systems*)

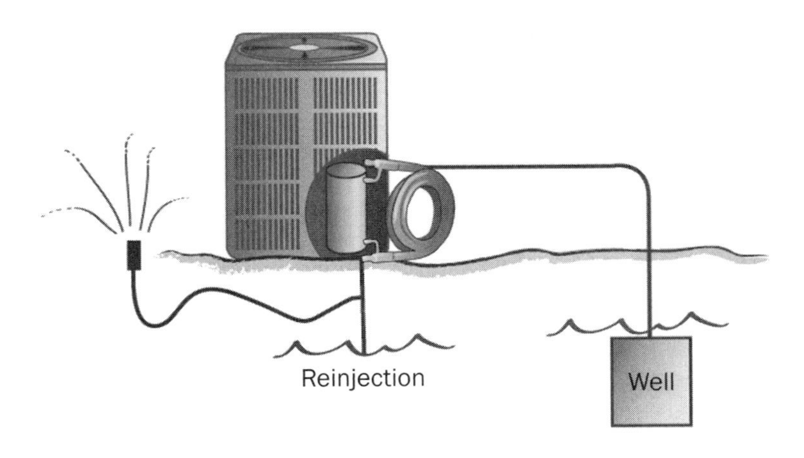

Figure 1-2 The conversion of an air-cooled air-conditioning system to a water-cooled or geothermal air-conditioning system prompted Jay Egg to found a company on renewable energy technologies. (*Sarah Cheney/Egg Systems*)

told me about Dr. Jim Bose from Oklahoma State University. Dr. Bose is recognized as the father of closed-loop geothermal heat pump technology. He is the founder of the International Ground Source Heat Pump Association (IGSHPA). Notice that the name says nothing about "geothermal" because this is truly a "ground-coupled" (or earth-coupled) technology.

I went to Stillwater, Oklahoma, and to Oklahoma State University shortly thereafter and became certified by IGSHPA as an installer. I learned about earth-coupled exchange through the use of polyethylene plastic piping.

Important Note: *Although no installer should be trusted completely, no matter how much one believes he or she knows, it is vital to verify proper credentials.*

In researching this book, we talked to many geothermal installers and experts around the world (Fig. 1-3), many of whom echoed this caution. Ted Chittem of Air Perfect, Inc., a geothermal installer in Milford, Connecticut, told us,

> It is not hyperbole for me to say that at least 85% of systems being installed are being installed poorly, and don't even meet the minimum specs of the manufacturer. . . . All the nice data that people read on brochures can be meaningless, because that might only hold true in a lab. In the real world, you have guys whose idea of charging a system is putting their hands on the pipe, and when it feels like a Budweiser they're done.

Figure 1-3 The gorgeous South Farm Homes in Hinesburg, Vermont, by Reiss Building & Renovation achieve near net-zero energy status thanks to geothermal HVAC, solar thermal water heating, efficient lighting, good insulation, and other measures. (*Photo by Brian Clark Howard*)

Even after being trained, I've made more mistakes than I can fit in this book. But this certainly doesn't mean that the benefits aren't worth taking the plunge—it just means that you need to exercise some caution. And keep reading this book!

By the way, it turns out that the first known use of geothermal heating and cooling was actually back in 1912, according to a Swiss patent. Groundwater *open-loop* (see Chap. 4) heat pumps have been in use since at least the 1930s, and they received a considerable amount of study in the 1940s and 1950s by the Edison Electric Institute. Still, there were a number of factors that limited the growth of the technology, including poor equipment quality. With the introduction of plastic pipe in the 1970s, and the work of Dr. Bose, geothermal heating and cooling entered the modern age and has been growing by leaps and bounds ever since.

What Is Geothermal?

The word *geothermal* has two parts: *geo*, meaning earth, and *thermal*, meaning heat. Thus, geothermal concerns using heat from the Earth. However, there are a few different applications of geothermal technology.

The heat of the Earth can be used to generate electricity, which is typically done on the scale of power plants (Fig. 1-4). When water is

FIGURE 1-4 Costa Rica, which has several active volcanoes, gets part of its electricity from geothermal power stations like this one. (*Photo by Brian Clark Howard*)

introduced to extremely hot rocks, it forms steam, which can be used to spin a turbine and generate electricity, as in a hydroelectric power plant or wind turbine. Traditionally, geothermal power plants have been established in the relatively rare regions where volcanic activity is near the Earth's surface, as in much of Iceland and parts of New Zealand. The government of Indonesia recently announced that it plans to invest significant resources in exploring volcano-based geothermal energy in the island archipelago. In all these situations, power can be generated relatively cheaply, as well as cleanly, with essentially no carbon emissions.

More recently, engineers have also been experimenting with geothermal power generation in other areas by drilling extremely deep wells, then pumping fluids in to superheat them. Unfortunately, this approach has had some technical problems, in addition to the great expense of deep drilling. Experimental projects in Europe and California have been blamed for causing small-scale earthquakes, which nearby residents have claimed damaged their property.

Still, many energy experts are bullish on geothermal and expect it to grow over the next few years. Today, there are more than 8900 megawatts (MW) of utility-scale geothermal capacity in 24 countries, enough electricity to meet the annual needs of 12 million typical U.S. households, according to the Geothermal Energy Association. Geothermal plants produce 25% or more of the electricity in the Philippines, El Salvador, and Iceland. The countries with the highest per capita use are Iceland, Sweden, and Norway, although the United States has the most total geothermal capacity, with more than 3000 MW. Eight percent of that capacity is thanks to California, which has more than 40 geothermal plants that provide about 5% of the Golden State's electricity, according to the Union of Concerned Scientists.

Geothermal electricity generation is an exciting field, but it is not what this book is about.

What Is "Earth-Coupled" Heating and Cooling?

Instead, this book is about shallow geothermal technology that is used to control the climate inside buildings—also known as heating and cooling. "Shallow," for our purposes, means no more than 500 feet below the surface. Most geothermal climate systems do not go below this depth, although they shouldn't be less than six feet below the surface either.

As I learned in Kentucky, even shallow caves maintain a fairly constant temperature year-round. In fact, some "green" architects are designing homes and buildings that are at least partially subterranean, to take advantage of that stable temperature and reduce the need for mechanical HVAC. However, there are also reasons why we don't build everything underground. That would be expensive and difficult, and people like to see out through their windows. Some innovative green designs get around this last problem at least partially with well-placed portals, skylights, and even sun reflectors called *solar tubes*. But for the most part, we are unlikely to radically change the basic design of our buildings. It's much easier to install a geothermal heating and cooling system in a new design of a type that we are already familiar with, or retrofit an existing structure—which we can readily do.

The geothermal AC concept works like this: We take the largely constant temperature of the earth beneath our feet and start from that point to begin heating or cooling our home or business. Because we are not using an outside fan, which passes already hot summer air over a coil that is sitting in the summer sun, we are saving energy. How much energy is determined by the difference between that outside air and the ground temperature. This difference could be between 10 and 25 degrees in summer (see Fig. 1-5*a*) and up to 50 or more degrees in the winter (see Fig. 1-5*b*). These basic factors, combined with other variables, give us *free energy*. Up to 80% of the cooling and heating needs of a building can readily come free from the constant temperature of the earth. We're just moving it. (By the way, the reason we say 80% is because the electricity needed to run the geothermal system typically totals around 20% of the cost it would otherwise take to heat or cool

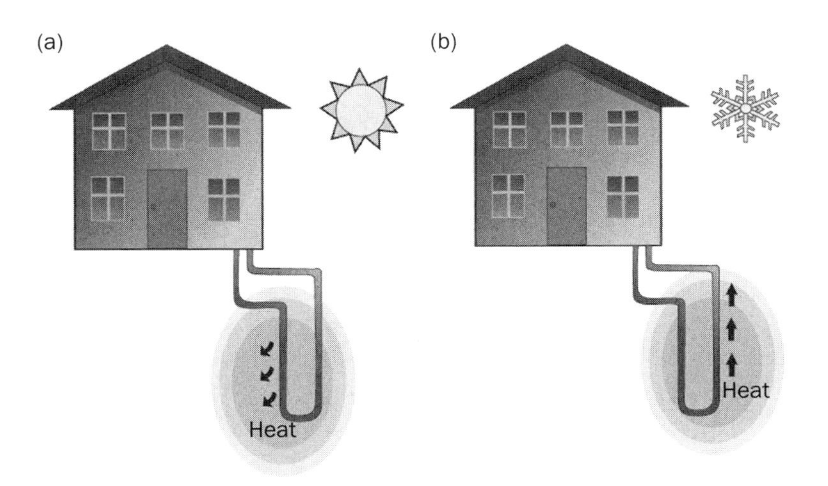

FIGURE 1-5 Geothermal cooling and heating uses the relatively stable temperature of the Earth to help heat a building in the winter and cool it in the summer. (*Sarah Cheney/Egg Systems*)

the structure with fossil fuels, electrical resistance heat, or air-sourced cooling.)

We call this an *earth-coupled* or ground source system because it uses the surrounding ground as a heat sink and a heat storage medium. An earth-coupled HVAC system also typically encompasses all the components of standard indoor climate control: heating, cooling, humidity control, zoning, air quality, air changes, and so on. For the bulk of this book we will simply call inside comfort conditioning either *geothermal air-conditioning* or *geothermal AC*, as these terms are quickly identified with the technology (Fig. 1-6).

Let's take a quick look at a few examples. Near Oslo, Norway, Nydalen Energisentral produces 50% of the energy needed in Nydalen's business and residential area through geothermal heating and cooling. The system services a college, hotel, several apartment blocks, and commercial buildings, through the use of 180 wells that are 200 m deep. The associated heat pumps produce 9.5 MW of heat and 7 MW of cooling, saving roughly 3000 tons of carbon dioxide emissions a year.

Also in Norway, Akershus University Hospital has 228 wells that produce 8 MW of cooling and 28 MW of heating. The country's national football arena, Ullevål Stadium, has 120 geothermal wells,

FIGURE 1-6 A look at the geothermal heat pump in the South Farm Homes in Vermont. The system is a water-to-water, earth-coupled design with fan coils. That is, water is used as the medium to heat or cool the air, instead of a direct exchange system. (*Photo by Brian Clark Howard*)

while the Oslo airport Gardermoen has 8 wells, which produce 8 MW. The airport investment paid for itself in just 2 years. A reclaimed paper mill in Drammen that is now home to offices, classrooms, labs, and a library has a 6-well geothermal HVAC system that saves 710,000 kWh annually, and is expected to pay for itself after 4 years, according to Nordic Energy Solutions.

The Clean, Green Energy with Great Promise

Geothermal heating and cooling is not well known by the general public, even though it has several key advantages and is steadily gaining in popularity. In fact, more than 1 million earth-coupled heat pumps have been deployed in the United States, according to the Stella Group, which is a marketing firm for alternative energy. About half of these systems serve residential customers and half serve commercial, institutional, and government facilities. Each year, American homeowners now install more than 50,000 geothermal heat pumps. Stella estimates the total market for geothermal HVAC in the United States at 3.7 billion dollars for 2009, including equipment and installation costs (and not reduced by government or other incentives, which you can learn about in Chap. 8).

Priority Metrics Group predicts a growth rate of 32% for geothermal AC to continue for the next few years, with the market exceeding $10 billion by 2013. Yet few people know how it works, or that it can be installed in nearly every location, for both new and existing construction.

There are plenty of good reasons to invest in a geothermal system, as you will soon see. An earth-coupled system is exceedingly reliable, quiet, and efficient. There is no smoke or fumes created, as nothing is combusted. A geothermal system provides steady, even temperature and humidity control throughout the day and night, without the extreme blasts of hot or cold air associated with conventional equipment. It will save you money over time—generally a good deal faster than competing technologies such as solar or wind. And it packs considerable environmental benefits.

Reducing Peak Demand

"Utilities should support geothermal HVAC to deal with peak summer demand. There is no better tool to lower their demand costs than geothermal," argues Ted Chittem of Air Perfect. This is partly because, of all on-site renewable energy technologies—solar, wind, tide, and the like—geothermal systems have the shortest payoff periods, typically only a few years (see Chap. 10 to calculate your own payback period). Geothermal units can be installed in almost any location, unlike wind, tide, or microhydro (small-scale hydropower), and they aren't dependent on weather or cloud cover. If more people switch to geothermal systems, it will decrease enough peak demand to prevent the building of new power plants and new transmission lines—both of which have major environmental, and public health, impacts. Otherwise, rising population and increasing demand will inevitably force such development.

In fact, according to the U.S. Department of Energy (DOE), a 400-MW natural gas turbine generator could be taken offline for every 99,500 homes or 1400 buildings of 100,000 square feet converted to geothermal heat pumps.

For commercial building managers, a geothermal system can result in real savings through *demand side management*. Since large operations use a lot of energy, utilities often assess these users with fees in proportion to their demand to help them cover the costs of buying expensive peak energy on the market. Take the example of a 200,000-square-foot office building in Tampa that reduced its demand by 175 kW as a result of installing a geothermal system. This customer had been paying $8500 in a demand "fee" before they began to pay for the first dollar for energy they actually consumed. But after the geothermal install that fee dropped to under $6800 for the same month.

Reducing the Use of Fossil Fuels

As we can see from Fig. 1-7, almost 50% of the energy used in the United States is generated by burning coal. Yet coal mining is dangerous and destructive, resulting in loss of life from accidents and loss of property from land subsidence, not to mention permanent scarring of entire landscapes through "mountaintop removal" mining techniques. The fuel produces notoriously dirty emissions and contributes to smog and poor air quality, which aggravates asthma and results in thousands of deaths a year in the United States alone, according to the Environmental Protection Agency. Coal pollutants create acid rain, lead to mercury poisoning, and are one of the biggest contributors to global warming.

In the United States, we get about 22% of our energy from natural gas and 19% from nuclear power, while all renewable energy sources combined account for only about 8%, according to the Energy Information Administration (EIA). Of that 8%, the biggest contributors are hydropower, biomass combustion, and wind.

According to the EIA, about 37% of the United States' total electricity use comes from the residential sector, while 36% comes from the commercial sector, and the balance comes from manufacturing. If we break down the residential use, as in Fig. 1-8, we see that heating and cooling account for about 31% of home electricity use, while kitchen appliances follow close behind with 27%. Heating, water, and lighting each make up 9%, followed by home electronics and laundry

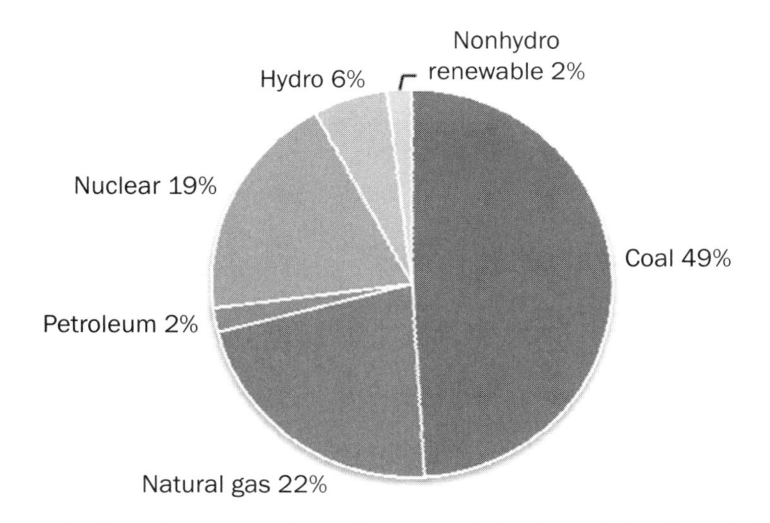

FIGURE 1-7 Fossil fuels such as coal and petroleum products account for about 75% of our energy use in the United States. (*Credit: Pew Center on Global Climate Change*)

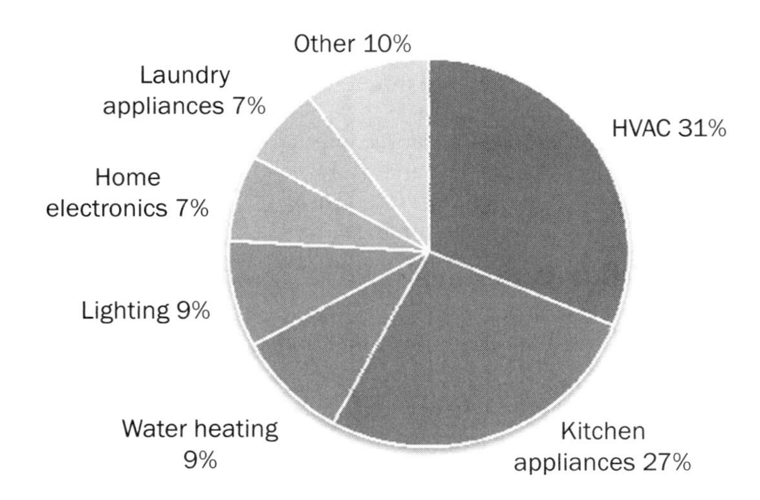

FIGURE 1-8 HVAC and water heating account for 40% of typical household energy consumption. Geothermal reduces this energy use significantly. (*Credit: Pew Center on Global Climate Change*)

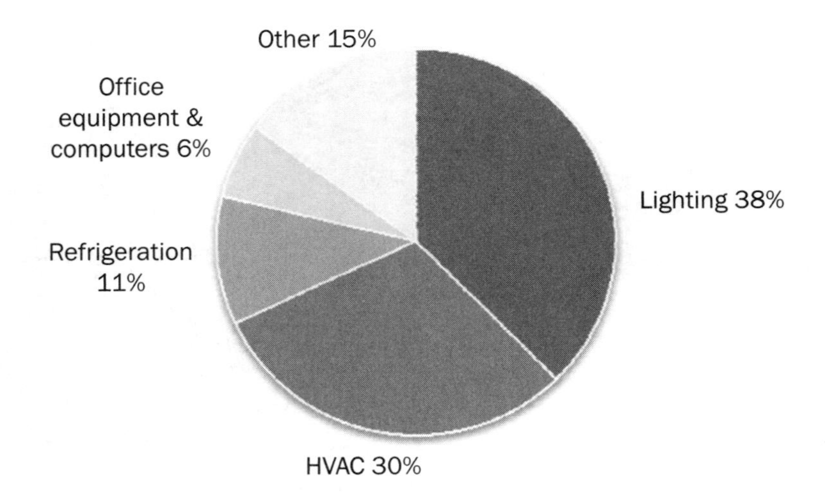

FIGURE 1-9 HVAC, refrigeration, and water heating account for 41% of typical business energy consumption. Geothermal reduces this energy use significantly. (*Credit: Pew Center on Global Climate Change*)

appliances at 7% each. The average American household spends $1900 a year on utility bills for heating, hot water, and electricity, according to the DOE.

On the commercial side (Fig. 1-9), lighting takes the biggest piece of the electric pie, 38%, while HVAC is 30%, refrigeration is 11%, and office equipment and computers account for 6%. Clearly, it takes a tremendous amount of energy to heat and cool our buildings, which today comes largely from fossil fuels.

Ted Chittem told us that many of his customers choose geothermal because they want to decrease their reliance on fossil fuels. "It's going to take a lot more dollars to buy oil on the international market in the future," Chittem told us. "Most of us who study this believe oil production has peaked, and as the world comes back from this economic downturn, and as a third of the people in the world, in India and China, start taking warm showers and driving $2000 cars, there is going to be a huge increase in energy prices."

Reducing Carbon Emissions

The electricity generated for a typical American home in a year puts out more carbon dioxide than two average cars over the same time period. If you add up home appliances, heating, and lighting, then

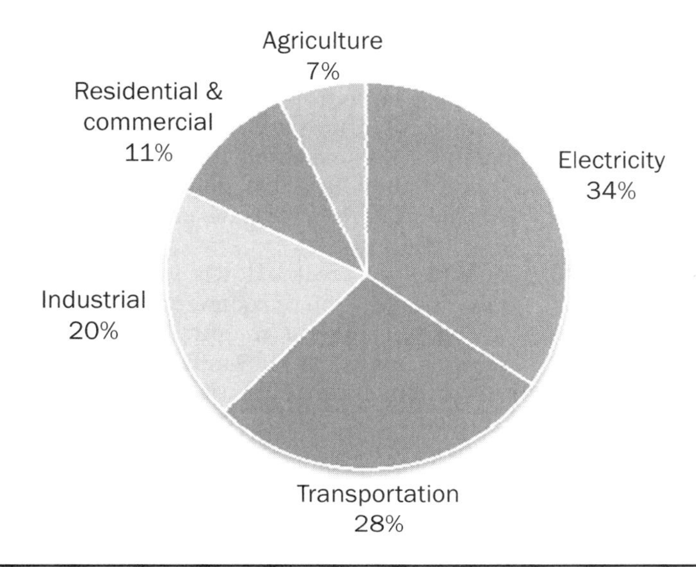

FIGURE 1-10 Carbon emissions by sector in the United States. (*Credit: Pew Center on Global Climate Change*)

you've accounted for almost one-third of U.S. carbon emissions (28%), which is more than the transportation sector (see Fig. 1-10).

This breakdown is especially significant because the United States accounts for one-fifth of total global greenhouse gas emissions from heating, cooling, and electricity. China accounts for slightly more than the United States, while the European Union and India are the next largest contributors. Unfortunately, the global trend is to increase emissions, despite rising awareness and international efforts to address the looming threat of climate change. Between 1990 and 2006, greenhouse gas emissions from the U.S. electricity sector grew an average of 1.5% per year.

By the way, in case you are wondering why experts refer to "carbon" as seemingly synonymous with greenhouse gas emissions, it's because carbon dioxide makes up roughly 98% of the greenhouse gas emissions from the electric power sector (coal plants account for more than 80% of the releases from that industry).

The good news is that a typical 3-ton (10.5 kW) residential geothermal HVAC system produces an average of 1 pound (0.45 kg) less carbon dioxide per hour of use than a conventional system. Therefore, if 100,000 homes converted to earth-coupled systems, the country could reduce CO_2 emissions by 880 million pounds (39,916,096 kg).

Take the example of one of the world's largest geothermal AC projects: installation of heat pumps in 4003 homes in the U.S. Army's Fort Polk in Louisiana. According to a report by Oak Ridge National Laboratory, this project reduced electrical consumption by 26 million kWh (33%), while eliminating consumption of 260,000 therms (27,429 MJ) of natural gas. Peak demand was reportedly reduced by 7.5 MW (43%) and carbon dioxide emissions were reduced by 22,400 tons (20,320,922 kg) per year.

When you break it down, 1.14 kW of peak demand was reduced by every installed ton of cooling, or 285 kW for every 100,000 square feet (9290 m^2) of building (roughly 71 homes). The Fort Polk geothermal retrofit was funded by $18.9 million in private capital, with cost savings to be shared by the investors and the U.S. Army over the life of the 20-year contract.

In recognition of their powerful ability to decrease environmental impacts, geothermal HVAC equipment and related controls can help owners earn up to 73% of the points needed to attain a coveted LEED (Leadership in Energy and Environmental Design) certification as a green building.

Coal-Fired Electricity for Geothermal versus Natural Gas

Despite the many environmental benefits of earth-coupled systems, critics often argue that running a furnace on natural gas could have a lower carbon footprint than a geothermal air conditioner—if the power plant supplying the small amount of electricity needed for the latter is coal fired. This argument is flawed for several reasons:

- The numbers don't hold up—in other words, it takes so little electricity to run geothermal HVAC that it pales in comparison to carbon emissions from traditional equipment.
- The relatively small amount of electricity that is needed for geothermal systems is likely to be replaced or offset with more renewable technologies in the foreseeable future. Strictly gas- or oil-fired heating, on the other hand, only adds to our dependence on fossil fuels.
- The cost of heating with gas or oil is up to five times higher than heating with geothermal. That's not sustainable, or fiscally responsible.
- Geothermal equipment can take advantage of load sharing (see Chap. 5), allowing a substantial reduction in total energy use, including savings for water- and pool-heating, coolers, freezers, ice makers, and much more.

Summary

In this chapter, we presented an overview of how geothermal HVAC technology can help us take advantage of the tremendous resource beneath our feet to heat and cool our spaces. We saw that roughly 80% of the energy needed for a building's climate control can be harvested from the surrounding soil. We learned that geothermal is a "green" renewable technology that can reduce our environmental impact and help us save money on lowered bills. Geothermal HVAC is a highly effective way to reduce peak demand for energy and lessen our reliance on fossil fuels.

CHAPTER 2

Heat Transfer and HVAC Basics

I was 10 years old when I first became interested in air-conditioning. I was standing in the living room of our house on Route 66 in Barstow, my sweaty face in the cold air stream of a "Sears Best" room air conditioner. The icy, cold air coming through the dark brown plastic louvers fascinated me. I could not figure out what process created the cold air. By contrast, heat from our gas furnace was not any harder to understand than heat from our Franklin stove.

I finally asked Dad, and he explained that the heat was removed from the air stream, leaving the cold air that I felt on my face. I remember thinking, "Yeah, right, like that makes any sense. . . . How is a machine going to take the hot part of the air away and leave only cold air?" I remember thinking that it was not exactly like separating pebbles from sand with a screen. I held my hands out in front of me, trying to imagine the air molecules in them looking different, the hot from the cold. I looked back at that air conditioner. And then I did what came naturally to the son of a high school English teacher: I read about it.

It all has to do with the spacing and speed of the molecules of air. When the excited molecules pass by a cooler surface, the temperature difference allows an exchange of energy from the warmer to the cooler medium. Thus, the refrigerant in the cool evaporator coil absorbs the heat from the air stream.

Understanding Heat Transfer

My family and I enjoy swimming in the natural springs of Florida. Here, the water comes out of the ground at 72 degrees. When I get into 72-degree water, it feels more than cool; it feels like ice water! When I walk into a 72-degree room, it is comfortable. Why is there such a difference when my body is surrounded by the same temperature either way? The answer is explained by the principles of thermodynamics.

Let's begin by looking at the three ways to transfer heat energy: convection, conduction, and radiation.

1. *Convective heat transfer* (see Fig. 2-1) results from the bulk movement of molecules within fluids, which include liquids and gases. When you feel warm air coming out of a supply grille, what you are feeling is convective heat transfer. Eventually, of course, the warm air will lose its heat and go back into the ductwork through a return air grille (where it will pick up more heat during the next pass through the air handler).
2. *Conductive heat transfer* (see Fig. 2-2) can take place in all forms of matter and involves transfer of thermal energy between neighboring

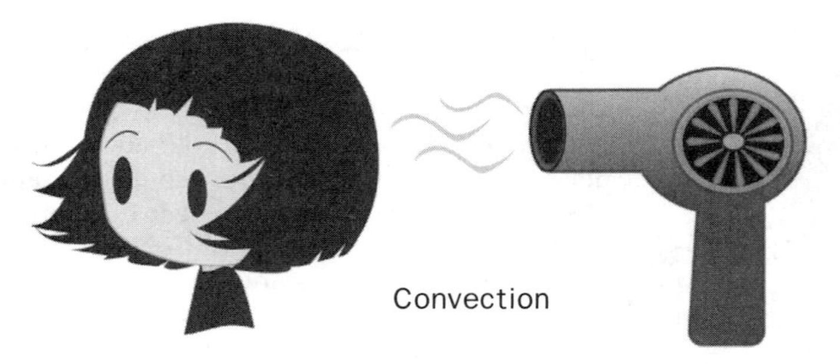

FIGURE 2-1 Using forced air to transfer heat, as in a hair dryer, is called convection. (*Sarah Cheney/Egg Systems*)

FIGURE 2-2 Like sodas in a cooler full of ice, conduction is an efficient way to cool or heat. (*Sarah Cheney/Egg Systems*)

molecules as a result of a temperature gradient. It always takes place from a region of higher temperature to a region of lower temperature, and acts to equalize temperature differences. Conduction does not require any bulk motion of matter. If you take a hot cup of cocoa and place a stainless steel dinner knife in it, heat will be effectively conducted within seconds. The same knife in equally warm air will take much longer to warm up, even several minutes. The same effect will happen with a dinner knife in a glass of ice water. The knife will rapidly give up heat and become cold as ice within a few seconds. Why? Because water is a much better conductor of heat than air (24 times better in fact) due to its far greater density.

As another example, most of us have put a warm can of soda into a chest cooler full of ice and water. Within minutes, it is ice cold through and through. Try that in an equally cool refrigerator and you will be waiting a couple of hours.

3. *Radiation* is heat transfer by electromagnetic waves, generally infrared for our purposes (see Fig. 2-3). This is independent of both convective and conductive processes. Radiant heat transfer can occur through the vacuum of space, through water, or through air. This is the warm feeling of the Sun on your face or the burning sensation you may feel when standing close to a heater with glowing red elements. This sensation is the same as the flash of heat you would feel if you were near an explosion. As soon as the source goes away, the heat is gone. The only reason that any of the heat from the Sun stays around at night is because of the thermal retention properties of the atmosphere, the Earth, and surrounding objects.

FIGURE 2-3 At a campfire, the girl's face and hands are warmed by radiant heat. Her back won't be warmed until she turns around. (*Sarah Cheney/Egg Systems*)

Heat Transfer and Geothermal HVAC

Geothermal air-conditioning systems use mainly the first two types of heat transfer: convective and conductive. The convective portion involves the duct systems, which consist of a series of return and supply vents. The finned coils of the system are designed to efficiently give up or take in heat from the air stream (see Fig. 2-4).

The conductive portion of the system has one or two parts or stages, and this portion is a significant factor that results in the high efficiency of geothermal air-conditioning. In a closed-loop, ground-coupled application, the liquid in the underground pipes (typically water with a little ethylene glycol antifreeze) conducts heat to and from the earth. The liquid in the pipes then typically goes through a water-to-refrigerant heat exchanger and conducts heat to or from the refrigerant, as in Fig. 2-5. The remainder of the process is completed by the refrigeration system.

In some cases, rather than a closed-loop system of pipes, groundwater is pumped from the earth and runs directly through the water-to-refrigerant exchanger, and then is injected back into the same water source, as in Fig. 2-6. Groundwater can also be pumped through a secondary exchanger that is connected to a closed loop serving one or more systems.

Conductive heat transfer through a liquid increases the efficiency of an air-conditioning system, just as stepping into a 72-degree spring feels like ice water compared to a 72-degree living room. The reason is simply the speed at which the heat transfer occurs. Remember the example of the knife? Water = fast heat transfer; air = slow heat transfer.

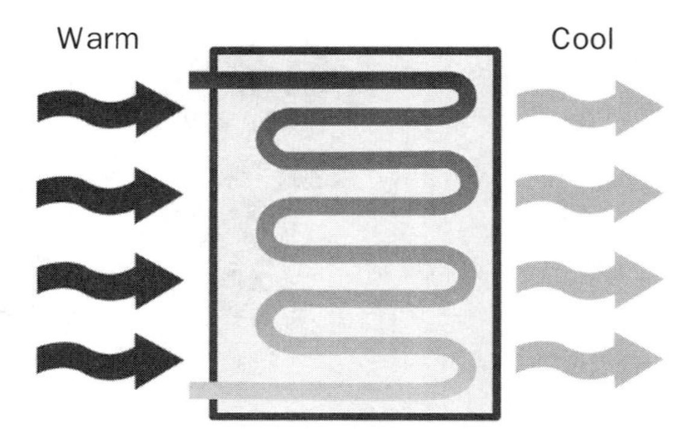

FIGURE 2-4 This refrigerant coil allows air to pass through the fins, where it is cooled in the process. This is convective heat transfer. (*Sarah Cheney/Egg Systems*)

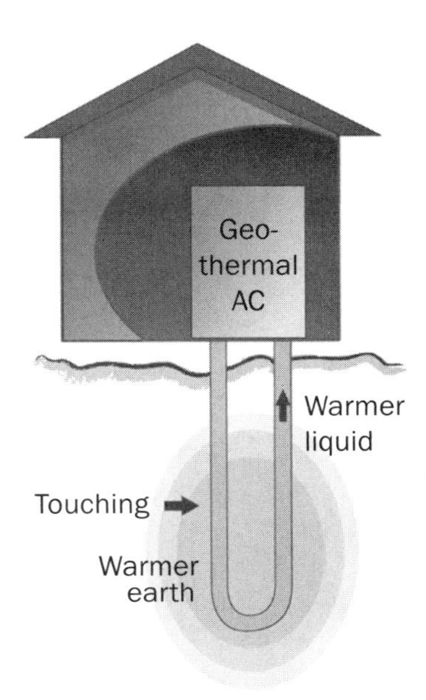

FIGURE 2-5 In this geothermal system, water in the pipes uses conductive heat transfer to pick up heat from, or dissipate it back to, the surrounding earth. (*Sarah Cheney/Egg Systems*)

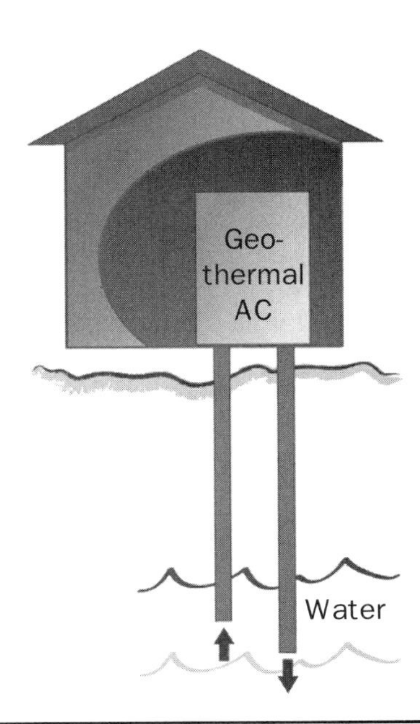

FIGURE 2-6 This is an example of an open-loop geothermal system, using groundwater to transfer heat. This is much the same as a closed-loop system, except in that case the same water stays within the pipes, as shown in Fig. 2-5. (*Sarah Cheney/Egg Systems*)

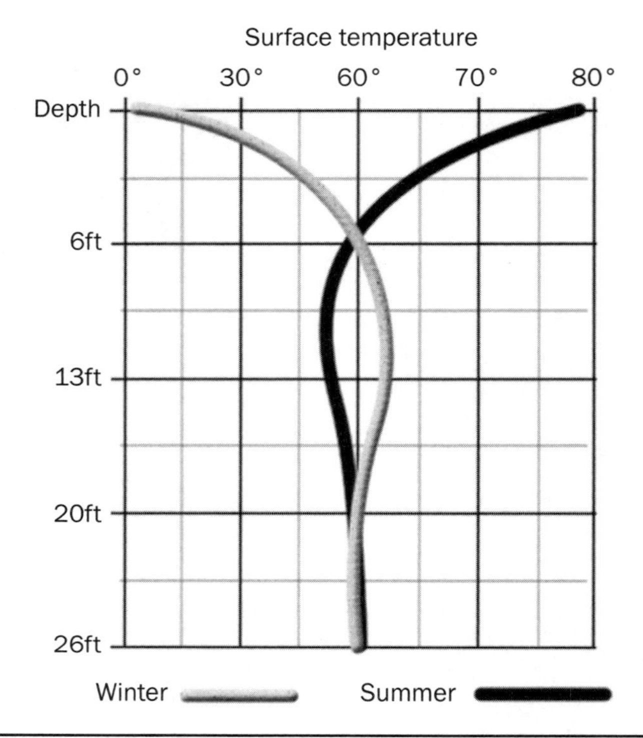

FIGURE 2-7 In most areas, 6-feet deep is the minimum depth that pipes should be buried. To take full advantage of the earth's stable underground temperatures, one must go about 26 feet deep. (*Sarah Cheney/Egg Systems*)

You can see in Fig. 2-7 that at about 26 feet deep, the temperature of the ground is nearly completely stable, while at 6 feet, the temperature varies a few degrees seasonally. Temperatures truly start to normalize at about 12 feet deep. If you look at the map in Fig. 2-8, you will see the range of temperatures in different regions of the United States. You will notice that the temperatures range from a low of about 37°F to a maximum of 72°F. Notice how mild this temperature range is, and this is roughly the same range from the extremes of Northern Canada to Central America. In fact, it is worth pointing out that we talked to a geothermal installer in Alaska, who told us that the technology can work great "up there," as long as the cooler earth is taken into consideration.

Another element of heat transfer that can work to the benefit of geothermal systems is a process called *thermal storage*. To understand this concept, ask yourself: If you just finished cooling your home after a nice hot summer, where did you put all that heat? Clearly, it has gone into

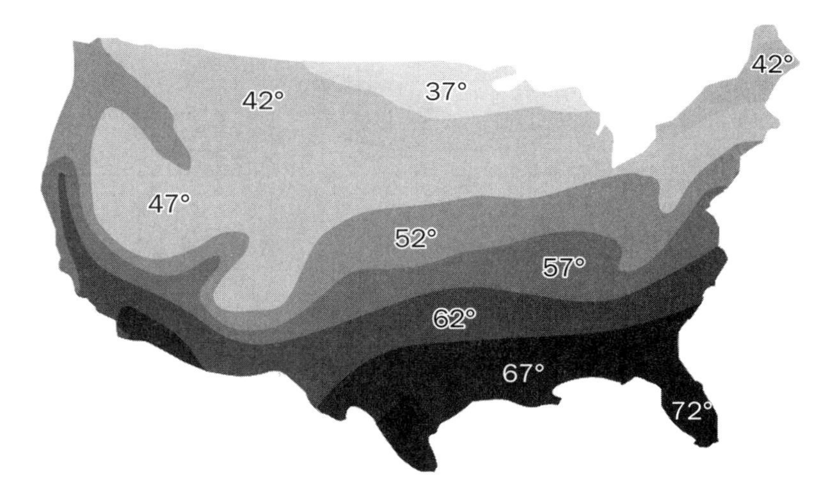

FIGURE 2-8 The standard mean ground temperatures indicated in this map give a clear picture of how well geothermal works. Even when air temperatures are below freezing, much heat is available to a geothermal heat pump. (*Sarah Cheney/Egg Systems*)

the ground. And like a big mug of hot cocoa, it's going to take a while before all of that heat dissipates. It could be weeks or months. So when you switch over to heat for that first cold winter night, you get to pull that heat you put down there right back up. . . . We call that *thermal storage*, or *thermal retention*. The same thing happens when you switch back from winter to summer. . . . Only now the ground is a lot like a nice cold glass of lemonade. It'll normalize in a few weeks or months, but you get to cash in on the thermal retention on those first hot summer afternoons.

The Main Parts of a Geothermal System

Geothermal air-conditioning systems have two or three basic components: The *earth-coupled portion* (also known as the collector), the *load portion* (the object or medium to be heated or cooled), and usually a third portion that provides the remainder of the energy needed for heating or cooling, usually *mechanical refrigeration* (see Fig. 2-9).

- The *earth-coupled portion* can be any type of system that uses the earth as the source or sink for doing a job involving temperature change. This system could consist of underground ducts or pipes made of a material that can conduct heat, such as metal

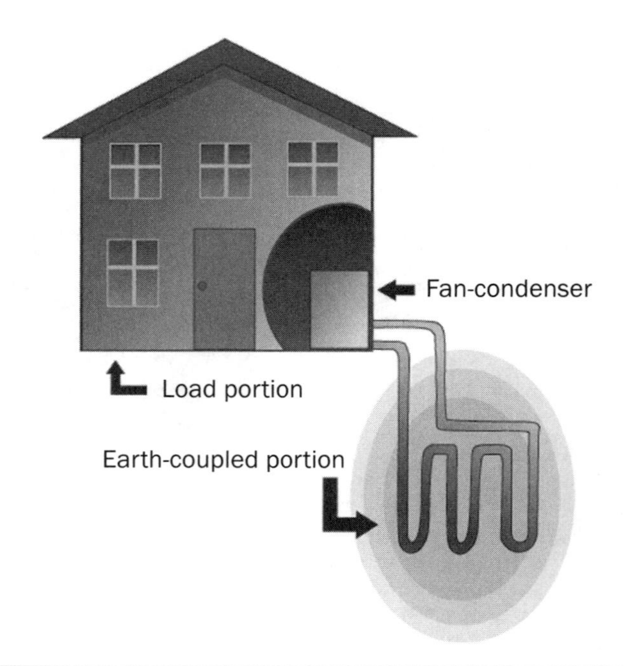

FIGURE 2-9 A geothermal air-conditioning system requires these main parts: (*a*) an earth-coupled portion, (*b*) a heat pump, (*c*) the building or object to be heated and cooled. (*Sarah Cheney/Egg Systems*)

or plastic. Or it could be either submersible or above-ground pumps designed to move groundwater.

- The *load portion* is the reason for any system. This load can be almost anything you want to heat or cool: a house, an elementary classroom, a commercial hot water tank, a drinking fountain, or a formerly snowbound driveway. To serve these loads, the installer can employ one of several different devices, such as forced air *hydronic exchangers*, refrigerant-to-water exchangers, or water-to-water exchangers. The choice is as limitless as our need for heating or cooling of almost any kind.

The Basics of Mechanical Refrigeration

To fully understand geothermal AC, it's also important to have a feel for the basics of mechanical refrigeration. There are four main components in most refrigeration systems: a condenser, a metering device, an evaporator, and a compressor (see Fig. 2-10). A liquid chemical

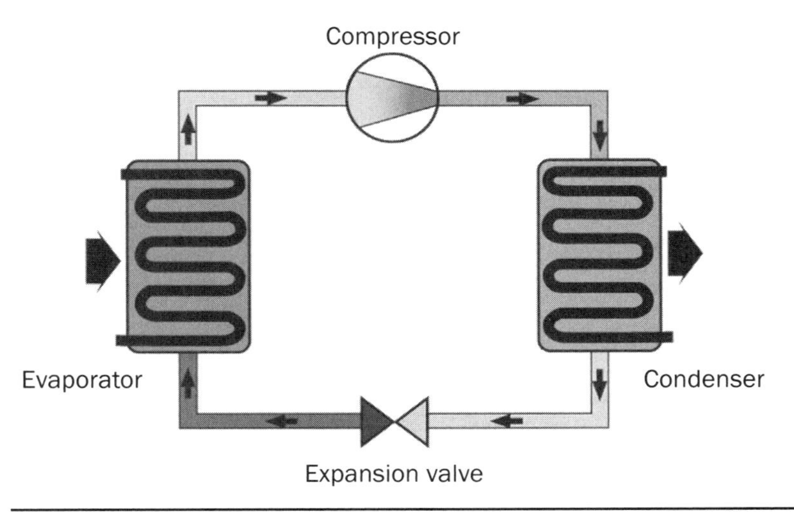

Figure 2-10 Most mechanical air-conditioning systems require four main components: (*a*) a compressor, (*b*) a condenser, (*c*) a metering device, (d) an evaporator. (*Sarah Cheney/Egg Systems*)

compound called *refrigerant* (most often a chemical called R-410A these days) is circulated within this closed refrigeration circuit. Refrigerant under high pressure and at a high temperature leaves the compressor and travels through the condenser, which does what it says: It cools and "condenses" the refrigerant to a liquid. As the unit's fan draws outside air across the warmer condenser coils, it results in cooling. Then the refrigerant passes through the metering device. Think of this as a spray nozzle on the end of a hose, as in Fig. 2-11. By the way, this concept sometimes requires a hands-on approach to understand completely, so please don't feel too badly if it doesn't completely make sense.

No matter how hot the water is in the hose, when you spray it out, it will be a cool mist after just a few inches of travel through the air. This cooling effect is called the *latent heat removal of vaporization*, or just *evaporation*. In the refrigeration circuit, the process is controlled, and the vapors are recovered and carried through the tubing of the evaporator to the compressor. The compressor takes the cool, low-pressure vapor, and recompresses it into a high-pressure vapor, increasing the temperature in the process. The reason for the increase in heat is primarily due to an increase in the number of molecules in the same space. The process continues as the high pressure, high temperature liquid again heads toward the condenser to do just that. . . . You know . . . condense! And that's it.

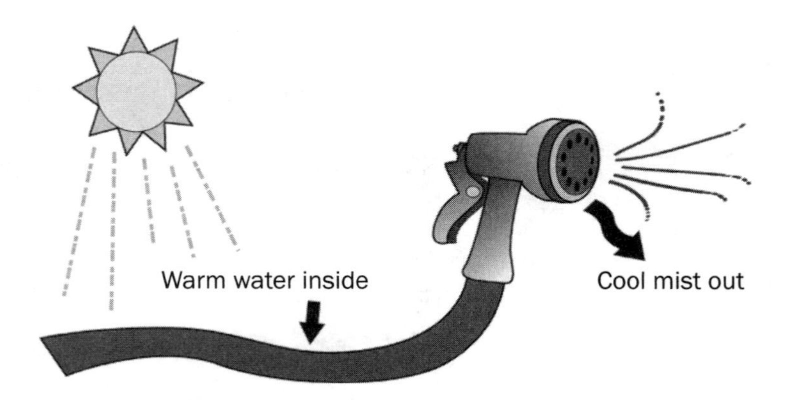

FIGURE 2-11 Evaporative cooling can be demonstrated by the example of 130°F water in a garden hose cooling almost instantly once it is sprayed in a mist from the nozzle. (*Sarah Cheney/Egg Systems*)

Types of Heating and Cooling Equipment

There are many different styles of air-conditioning and heating equipment. We will attempt to help identify the major systems, but note that there are numerous exceptions to every rule. While some unusual and custom-built systems work very well, others can be so complex that if the designer becomes unavailable, abandoning it for an industry standard may be more cost-effective than trying to keep it running.

In general, an air-conditioning and heating system can have split components, and can be packaged, cooling-tower sourced, solar-heat sourced, or passive (like a cave).

The indoor unit, usually called the *air handling unit*, can often be found in one of five places: a garage, closet, mechanical room, attic, subfloor, or above a drop ceiling between floors. They will typically be connected to a ductwork system, which channels air to the designated rooms of the structure.

A *split system* air conditioner can often be identified by an outdoor condensing unit. This outside unit will usually have a fan blowing upward out the top. These units have gained popularity over the last 30 years because they move the noisier compressor section outdoors. The coil and fan are housed inside in a box called an *air handler*. As you can see in Fig. 2-12, the two sections are connected by two refrigerant pipes.

A *package system* is by definition a system that has all of the components in one box. These are often referred to as *rooftop units* (RTUs) or *unitary* systems. They are commonly installed on rooftops of homes in certain regions of the country, or beside a house or mobile home.

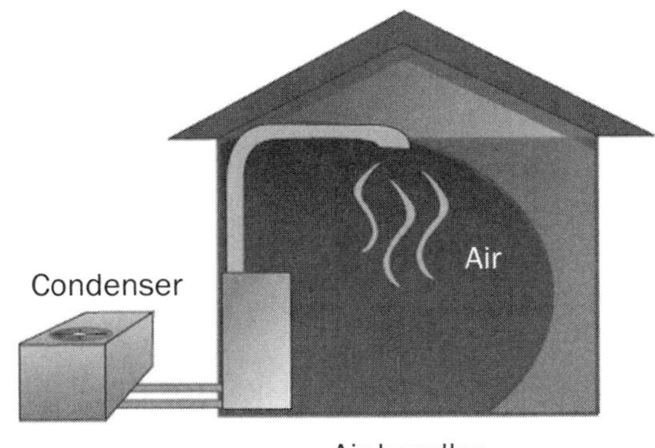

Figure 2-12 Most homes have split systems, which are distinguished by an outside condenser and an inside air handler connected by two refrigerant lines. (*Sarah Cheney/Egg Systems*)

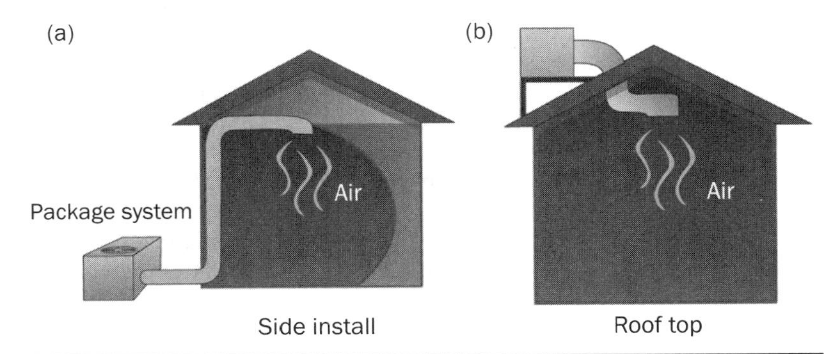

Figure 2-13 Packaged air conditioners can be installed (*a*) beside the structure or (*b*) on a rooftop, but must be outside of the structure they are cooling or heating. (*Sarah Cheney/Egg Systems*)

Figures 2–13*a* and 2–13*b* are two examples. It is typically considered favorable by many to buy a packaged system because there is no chance of installer error in the joining or sizing of the refrigerant tubing. On the other hand, these units must be installed outdoors, subjecting the entire unit, including portions of the air duct system, to weather and related extremes.

As you are probably aware, there are a number of different ways to power heating or cooling systems, such as by natural gas, oil, coal, or wood-fired furnaces; electrical resistance heat; heat pumps; passive solar; adiabatic coolers; direct expansion systems; chiller systems; absorption systems; and cooling towers. Let's take a closer look.

Conventional Fuel Burners

A gas- or oil-fired heater is commonly called a *furnace* and involves an exchanger or a burner that contains an open flame within an air handler. A small fan provides combustion air and forces the CO_2-laden exhaust out through a double-wall B vent to the outdoors. The open flame heats the exchanger and the fan in the air handler pulls air through the fingers of the burner/exchanger assembly. These can be part of a split system or a package unit.

The very best fossil-fuel furnaces achieve efficiencies above 90%, although note that they cannot exceed 100%, unlike geothermal systems, which have as part of their equation the "system" of the surrounding earth to draw from, versus just the energy you supply conventionally. Older, still functioning gas furnaces are often as bad as 50% efficient, meaning that half of the heat energy you paid for is simply wasted through your chimney or inefficient ductwork; it literally goes up in flames. Not only do newer furnaces have tighter tolerances, but they also employ additional fans and controls that coax the most utility out of the heat produced. Even so, they burn a nonrenewable (and toxic) resource, and since they can't draw on stored energy that is locally available, they can't possibly approach the efficiency of clean, safe geothermal HVAC.

Less common in North America these days, but remembered by our grandparents, and still widely in use in much of the world, are coal- and wood-fired heating systems. These can either be hydronic, radiant, or a combination of the two. If the coal or wood is combusted in a fireplace, it is typically radiant, which means the heat radiates from the device in which it is burned. Sometimes a blower is added to make the system more efficient by pulling more heat out of the air before it vents through a chimney or vent stack.

Sophisticated blowers are, in fact, used on new *pellet stoves*, which are becoming popular with back-to-the-landers and some environmentalists. Pellet stoves are extremely efficient in terms of coaxing heat out of wood material, and their high-temperature, controlled burn produces greatly reduced emissions and only minimal waste ash. Pellets are normally pressed from waste wood that would otherwise be discarded from timber mills, furniture factories, and the like. These stoves use a small amount of electricity to run fans and controls, not unlike geothermal HVAC.

If a furnace is placed outside of the space it is servicing, it is typically heating a store of water, and then is called a *boiler*. The water is pumped to radiators or fan coils throughout the building, providing heat as needed. This is called a *hydronic* system, regardless of the fuel.

Electric Resistance Heating

You may have lived in, or visited, a place with *electric heat*, also known as *electric resistance heating*. Electric heat can supply a centralized forced-air system, although this is less efficient than burning a fuel on-site, so this option often doesn't make sense. More commonly, electric heat is delivered through heaters in each room, either as electric wall, baseboard, or radiant heaters. Most portable *space heaters* operate by electric resistance heating.

Although this process converts nearly 100% of the electricity used to heat, when it comes to actually transmitting the electrons from a power plant, it becomes less efficient, and more costly than producing heat on-site. Therefore, electric heat results in higher heating bills in cold climates. It is better suited to mild locations, where it doesn't make sense to invest in more expensive HVAC equipment, or in temporary structures.

Heat Pumps

If electricity is the only choice available, the U.S. DOE suggests going with a heat pump over electrical resistance heat. At least for most climates, heat pumps generally cut electricity use by 50% over resistance heat. We'll get into heat pumps in much more detail in the next chapter, but suffice it to say that a heat pump is simply a device that moves heat from one location (the "source") to another location (the "sink" or "heat sink"), by using mechanical work.

Heat pumps can draw heat from the air, water, or ground. Air-source units (as in Fig. 2-14) are sometimes viewed skeptically in colder climates, since it's true that they often do not work well when temperatures fall below around 23°F (−5°C)—although it is possible to supplement them with a secondary source.

Passive Solar

One of the "greenest" ways to heat or cool a building is to use *passive solar* design to take advantage of the Sun's free energy, without needing additional inputs from fuels, the grid, or any other source. Ancient people knew this well, which is why they built their homes in caves, out of thick mud bricks, or with courtyards and overhangs. But in the last century or so, in an age of relatively cheap fossil fuels and "modern" design aesthetics, we have forgotten many building techniques that were more in harmony with the land.

We can't get into passive solar design in any detail here, but the good news is that more and more people are reawakening to the possibilities, and combing the latest in creature comforts and technology

FIGURE 2-14 A view of an air-source heat pump that can be used to heat or cool a building (though not nearly as efficiently as a ground-source heat pump). (*Egg Systems*)

with blueprints that save energy and add value. A few of the emerging techniques include planting shade trees, orienting the structure to take advantage of the seasonal march of the Sun across the sky, promoting cross-ventilation, installing awnings and overhangs, recessing windows, opening clerestory and soffit vents, building with the contours of the countryside, and optimizing the natural "heat sinks" of materials. High-concept test homes can maintain comfortable temperatures year-round with infinitesimal utility bills. But even if you can't redesign from the ground up, almost everyone can employ at least a few passive solar ideas to reduce their energy footprint, not to mention their monthly bills.

Adiabatic, Evaporative, or Swamp Coolers

Adiabatic cooling is a process in which the temperature of a system (or a building) is reduced without directly adding or subtracting heat. (Often this is done by changing the pressure of a gas.) Therefore, adiabatic coolers, also known as *swamp coolers* or *evaporative coolers*, work without mechanically extracting heat. In a swamp cooler, a relatively small circulating pump picks up water from a sump and deposits it onto evaporation pads. A main blower then draws air over the pads, which results in cooling (see Fig. 2-15).

With a swamp cooler, it is not uncommon to see 100-degree outdoor air cooled down to 75 degrees, provided the climate has low enough

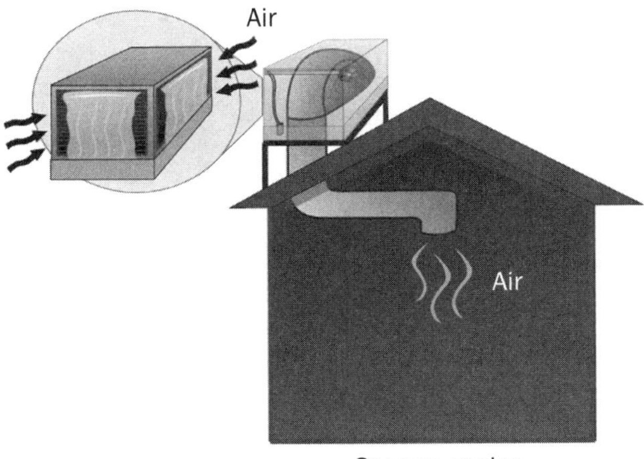

FIGURE 2-15 Swamp coolers are popular in dry climates. They can effectively cool a structure by humidifying the air. . . . This results in what is referred to as adiabatic cooling. In areas where the humidity level is already high, swamp coolers are relatively ineffective. (*Sarah Cheney/Egg Systems*)

relative humidity levels. The scientific reason for this can sound complicated, but these coolers are highly efficient in areas where the climate is dry. They don't require the compressor of standard refrigeration systems, which uses a great deal of energy. As a result, evaporative coolers typically use 75% less energy than traditional AC. And the coolers cost 50% less up front, owing to their simpler design. Swamp coolers also have fewer moving parts and require no ozone-damaging chemicals, although they typically do require more maintenance than standard AC. Residents of parts of the dry American West have used them effectively for decades, and many people can even get cash rebates from their utilities for installing qualifying models, since they help reduce peak summer loads.

One downside is that swamp coolers use a lot of water, typically between 3.5 and 10.5 gallons per hour. This can be a concern in water-stressed regions, although there can be ways to "recycle" some of the water used for other purposes, such as for irrigation. Swamp coolers also add humidity to the air, which can be undesirable. The air is not normally recirculated, but is discharged through barometric relief dampers, or through open windows, to avoid moisture buildup. (Interestingly, a Colorado-based company called Coolerado makes evaporative coolers with a second loop for heat exchange, so no humidity is added to outgoing air. However, the process is a bit less efficient, and the units are more expensive.)

Direct Expansion Systems

Direct expansion systems refer to refrigerant-based systems that have an evaporator or a condenser coil that is exposed directly to the final load or source (see Fig. 2-16). This is probably the most common type of cooling system you will find, certainly in residential applications, and often in commercial systems. The greatest advantage may be the low initial cost. A disadvantage is that the direct expansion outdoor coil, often called the *condenser section*, has its efficiency directly subject to extremes of outdoor conditions. This part is also often noisy and prone to wear and tear. The typical geothermal air-conditioning installation eliminates such outdoor equipment.

Chillers

Chillers are refrigeration systems that provide chilled water or some other "water-like" liquid as a medium to the final load (see Fig. 2-17). The chilled water can be used to cool a number of loads, including air coils, secondary exchangers, and processes. You will find this commonly in use in large commercial facilities with numerous air-handling units called *fan coils*. Because the coolant medium is typically water, these systems are often referred to as *hydronic* systems.

Hydronic systems are among the most efficient in regular use. This is due to several factors we have discussed, including the advantage of using liquids for heat transfer, and because there is no phase change (liquid to gas, or vice versa) to the final load, so the load can

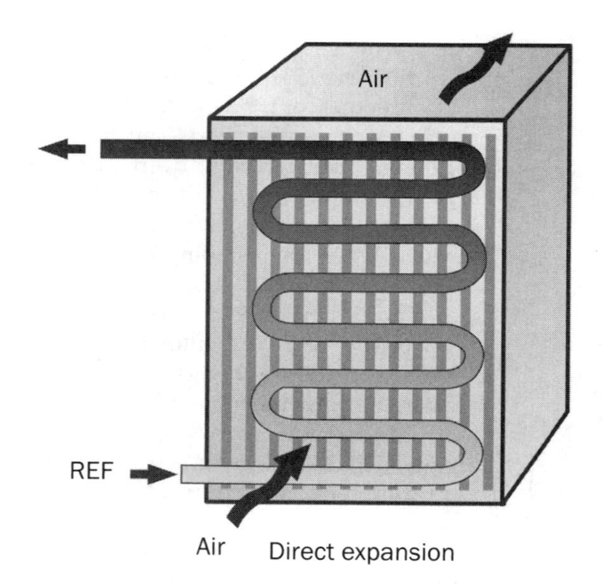

FIGURE 2-16 Direct expansion systems have an evaporator or condenser coil that is exposed to the final load or source. (*Sarah Cheney/Egg Systems*)

be divided up more easily. Chillers can be air-cooled, but are much more efficient when water-cooled, which may require a cooling tower outside of the building. Cooling towers can often be more efficiently replaced by a geothermal system, or at least supplemented with a geothermal source, as in Fig. 2-18. This will be expanded upon in subsequent chapters.

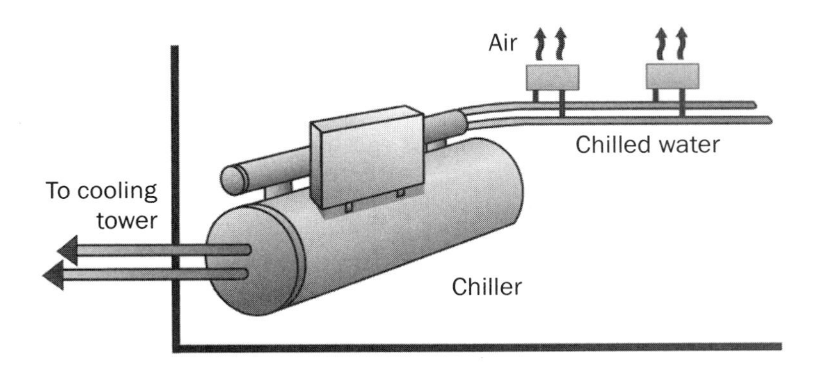

FIGURE 2-17 Chillers or hydronic systems provide chilled liquid to the final load, such as air handlers for an individual office space. (*Sarah Cheney/Egg Systems*)

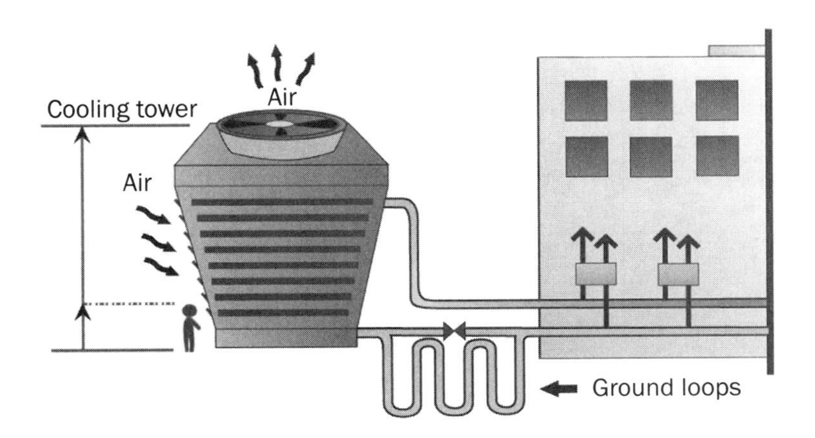

FIGURE 2-18 Larger commercial buildings often use cooling towers to provide coolant water to the chillers or heat pumps in the building. The cooling tower uses the process of evaporation to cool the water down to the "wet bulb" temperature. (*Sarah Cheney/Egg Systems*)

Absorption Chiller System

In an *absorption chiller system,* the process is similar in some ways to conventional electric vapor compression systems (standard refrigeration). Absorption cooling substitutes a generator and absorber, called a *thermal compressor,* for a typical electric compressor. Efficiency and lower operating costs are achieved through the use of a pump rather than a compressor and a heat exchanger to recover and supply heat to the generator.

A double-effect absorption cooling system adds a second generator and condenser to increase the refrigerant flow, and, by extension, the cooling effect, for a fraction of the heat needed for a single-effect system. The primary source of heat for these systems must be either waste process heat, such as low-pressure steam or hot water, or solar thermal heat. The waste heat from the absorption process can be removed through geothermal heat transfer versus the normal process of a cooling tower, reducing maintenance, chemical use and water consumption.

Cooling Towers

Cooling towers, like condensers, are installed outside of a building, normally in commercial settings, such as on hospitals or schools (see Fig. 2-19). Cooling towers can take on several different looks, and they have many applications, from nuclear power plants to coal-fired plants and factories. In HVAC, they are often part of a liquid-cooled refrigeration system and rely on the principle of evaporative cooling. The

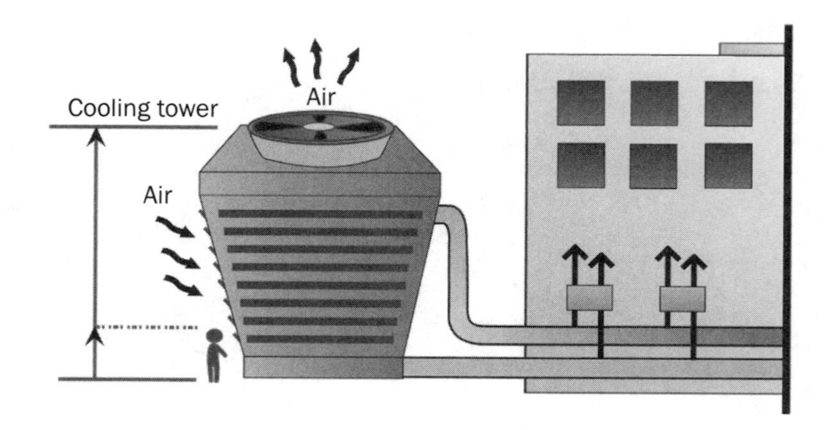

FIGURE 2-19 Cooling towers are normally large and lose considerable amounts of water to evaporation. They also require a considerable amount of chemical treatment to remove the minerals and impurities that build up in them. (*Sarah Cheney/Egg Systems*)

towers cool condenser water down to what is referred to as the *wet-bulb temperature*—defined as the lowest temperature that can be reached by the evaporation of water in a given setting. Wet-bulb temperature is affected by relative humidity, and the lower the wet-bulb temperature, the more efficient the cooling tower in summer heat. *Dry-bulb temperature*, in contrast, is independent of moisture and is what we usually think of as "air temperature" (it is measured by a dry thermometer that isn't exposed to radiation).

In the case of a HVAC cooling tower, water is cooled down, then sent back into the building to remove heat from the equipment it services.

The Critical Issue of Equipment Sizing

Sizing of air-conditioning and heating systems remains a mystery to many, although it is extremely important in terms of efficiency and comfort, not to mention price. You may notice that when asking a contractor a question such as, "How many tons of air do you need for an average home?" typically you will get a series of comments or objections to the query and statements like, "There is no way to determine that until they do a thorough home inspection and run all the calculations through the appropriate programs."

On the other hand, you may get an answer that works everywhere according to that particular individual, such as 500 square feet per ton in residential and 250 square feet per ton in commercial. The real answer is probably somewhere in the middle.

My first construction job as a geothermal contractor was for a 10,000-square-foot home in Carrollwood, Florida. I knew I was taking on a large project, but I had no idea how difficult it would become. During the permit application process, I was given many forms to complete. None of these would prove to be as involved as the Manual J Residential Heat Gain and Loss Analysis (see Chap. 6 for more details). In order to calculate the amount of cooling and heating needed for a structure, some of the items taken into account are the insulation values (*R values*) of each wall and ceiling, the square footage of windows and doors, the orientation of walls to the outdoors or nonconditioned indoor space, overhangs and distance of windows from the overhang. These measurements can be grueling and especially difficult without a computer program.

Luckily, I had some help. Henry Stobaugh was (and still is) the chief mechanical inspector for Hillsborough County, Florida. He gave me a few pointers to get through the process, since he took a personal interest in me and in geothermal air-conditioning. Henry was instrumental in getting me in the door with several agencies and companies that were pivotal in my early successes. He arranged for my first training classes in 1994, training Hillsborough County inspectors on earth-coupled technology. But I digress . . .

Back at that home in Carrollwood, I finished the permit forms and mechanical prints after considerable trial and error, and the permit was issued. Henry walked that first job with me during the final inspection. I can remember every detail of the job even now as I reflect, right down to the 7-day programmable thermostats, a hi-tech item for that time. Tampa Electric Company, the local utility, metered that job, and eventually approved amazingly generous rebates for that time; $850 per geothermal heat pump, no limits per home. In this case, $3400 was rebated. Not too bad, but it pales in comparison to the average $6000 to $10,000 tax credits currently given for geothermal systems (see Chap. 8 for details).

After the difficult exercise of manually calculating heat gains, I purchased a program from Wright Soft that aided in heat calculation. This proved to be even more difficult than the manual calculations the first time or two. And it doesn't get easier with each upgrade. In our office, we have engineers who use the program to do their calculations and lay out the ductwork and airflow. Without exception, you can hear the pain and suffering from each new upgrade. Often these expensive upgrades are necessitated by code changes, as in the most recent change from 10 to 13 SEER minimum standards.

Here are some rules of thumb. I like to say that smart people learn from their mistakes. Really smart people learn from others' mistakes. Hopefully, this will put you, the reader, in the really smart category. Since I have already made almost every mistake possible in my pursuit of geothermal air-conditioning, I will assume the role of those "other people" that we like to learn from. Now, we can get down to the business of sizing your air-conditioning and heating systems. Here are the basics:

In the United States, heating and cooling is measured in *British thermal units* (*Btu*). A Btu is equal to the energy it takes to heat up 1 lb of water 1°F. Oftentimes, gas heat is also measured in *therms*; one therm is 100,000 Btu. Air-conditioning is measured in Btu or *tons*, which correlate with tons of ice. Therefore, a ton of air-conditioning is equal to the amount of cooling that would be done by melting 1 ton of ice, as in Fig. 2-20. Lastly, a *kilowatt-hour* (kWh) is equal to 3412 Btu. Don't bother trying to remember all of that. . . . There won't be a test as far as I know. . . . This list is provided just so you know that there is a basis behind the terms. Here they are for quick reference:

- *Btu:* British thermal unit
- *ton:* A unit of heat removal equal to the removal of 12,000 Btu
- *therm:* a unit of heat equal to 100,000 Btu
- *kilowatt-hour:* a unit of energy equal to 3412 Btu

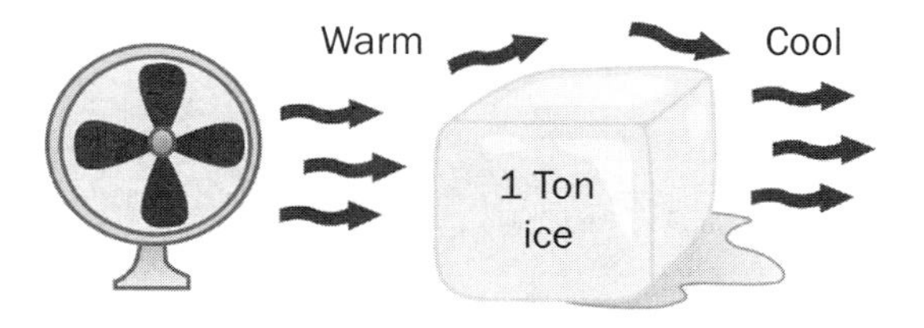

Figure 2-20 A "ton" of air-conditioning is the amount of cooling that would be produced by 1 ton of ice melting in front of a fan. This is equivalent to the removal of 12,000 Btu of energy. (*Sarah Cheney/Egg Systems*)

Important Note: Most HVAC and refrigeration calculations in the United States are related to Btu and Btuh (per hour). If you're reading this outside of the United States, you'll be using joules and watts as the basis of energy. One kilowatt-hour equals 3.6 mega Joules.

To get a rough idea for the size of system you may need to cool or heat a home of average construction, consult Fig. 2-21, which shows general ranges across the United States.

To cool a home of average construction, you will typically need: ____________________ (see Fig. 2-21).

To heat a home of average construction, you will typically need: ____________________ (see Fig. 2-21 depending on the maximum difference between conditioned air and the outdoor air temperature.)

What Is the Size of My Current System(s)?

This is another question that is often easy, but can become difficult with certain brands of HVAC equipment. Even after decades of experience, we still have to sometimes call in to the manufacturer with our model and serial number. Usually, the answer is embedded in the model number, but it can be difficult to extract until you have the cipher code. It's a good idea, once you get the model and serial number, to keep them handy for the times you will need to call for service.

You should see highlighted on the condenser unit the Btu in thousands, which can be converted to tons by simply dividing by 12. On the furnace or air handler you will find the heat capacity in kilowatts or Btu, depending on the fuel source. In the serial numbers, you will see that you can often find a date code. You simply need to know how to read it. For example, in the WaterFurnace model NDV049111CTL, you find the "049" and divide by 12: The answer is 4.08 tons. In the ClimateMaster model TTV064AGC01ALKS, you can see how easy it is to extract the "064" and divide by 12 to get an answer of 5.33 tons.

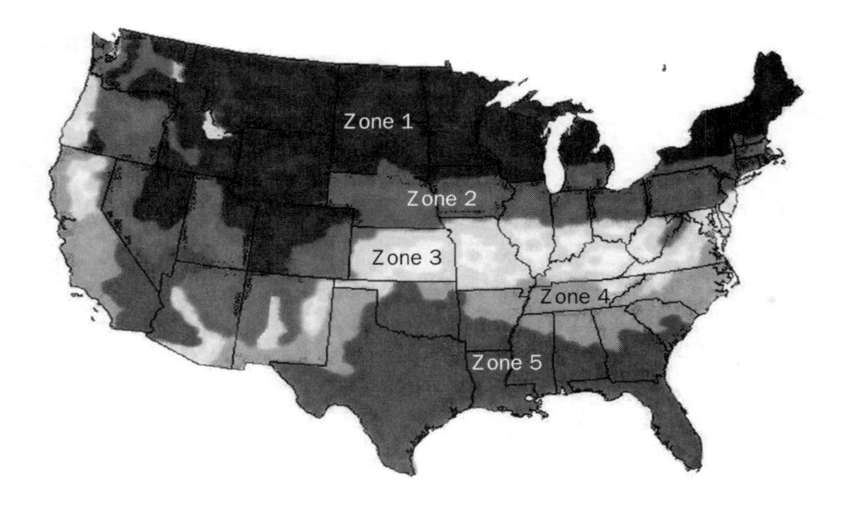

Air Conditioning Range Square Footage Per Ton
ZONE 1: 467 to 734
ZONE 2: 450 to 700
ZONE 3: 434 to 667
ZONE 4: 417 to 633
ZONE 5: 400 to 600

Heating Range BTUs per square foot
ZONE 1: 50 to 55
ZONE 2: 45 to 50
ZONE 3: 40 to 45
ZONE 4: 35 to 40
ZONE 5: 30 to 35

NOTE: These calculations are based on averages and are intended only as a general guideline. Individual calculations will vary depending on different factors such as types of windows, doors and insulation. For exact figures energy calculations should be completed by a certified professional.

FIGURE 2-21 General guidelines for sizing HVAC systems in the United States.

Summary

We have seen the basics of mechanical refrigeration, as it occurs in home refrigerators and conventional air conditioners. We have also learned the basic components of geothermal systems, including the earth-coupled portion and the load portion. We have also seen the basic types of heat transfer—convection, conduction, and radiation—and have seen how these affect system efficiency.

It's crucial to size systems properly, and you should have a heat gain analysis done to determine your actual cooling and heating load before you make your final decision. I worked with a builder and a homeowner on a nice little custom project in 1994, in which we installed a horizontal, slinky-type loop underneath a drain field (more on this later). We had completed calculations for the 2700-square-foot house, confirming the need for a 5-ton capacity geothermal air-conditioning system. After we had several meetings and disagreements on the capacity, R-values for the insulation, windows, roof color, and other factors, I acquiesced to their relentless insistence that this would be only a 4-ton load. The builder was nowhere to be found, figuratively speaking, during the next three brutal summers, each of which brought weekly phone calls and repeat visits to tweak this air flow, or check the ductwork one more time, or to better insulate the plenums.

So I finally did what we do when our customers aren't happy; you guessed it, I put in a new unit sized at 5 tons. I've not heard a peep of complaint since. I've also never allowed a system to be improperly sized since that time.

CHAPTER 3

Geothermal Heat Pumps and Their Uses

Near the southern tip of Manhattan, in the trendy, windswept neighborhood of Battery Park City, sits the 35-story, 400,000-square-foot Millennium Tower (see Fig. 3-1). The new building is home to 234 condominium and apartment units, and boasts a number of green features, including 932 ClimateMaster water-to-air heat pumps that are woven into the building's closed water loop, boiler, and cooling-tower climate control system.

The designers of the Millennium Tower estimate that the geothermal technology reduces the building's energy expenditures by about 22%, compared to a conventional system (even more energy would be saved if the cooling tower and boiler were eliminated in favor of a pure geothermal system). Two different types of heat pumps are used: a slim vertical style tucked away in closets in kitchens and living rooms and a console unit that fits snugly below bedroom windows (both visible in Fig. 3-2). Each model provides a ton to a ton-and-a-half of heating-and-cooling capacity.

Earth-coupled heat pumps are versatile and powerful devices that are gaining wider acceptance around the world. Here we take a look at some of their most important applications.

Passive and Forced-Air Earth-Coupled Duct Systems

Passive and forced-air earth-coupled duct systems are often considered to be Do-it-Yourselfer projects. Generally, a design is procured online for a particular structure. The contractor or homeowner secures a given footage of plastic pipe, several inches in diameter, and installs it in the relatively cooler or warmer soil. Air is forced through the piping from the outside or recirculated from the inside (see Fig. 3-3).

The benefits of these systems include low energy consumption, easy installation, and little training needed.

Some of the concerns include condensation on the inside surface of the pipes underground, which can lead to the formation of mold and bacteria. Where there is no mechanical refrigeration in use, there may

FIGURE 3-1 Millennium Tower in downtown New York City sees substantial energy savings thanks to geothermal HVAC. (*Photo by ClimateMaster*)

be discomfort from lack of cool air under higher summer loads and inadequate humidity removal. This again can lead to bacteria and mold growth, which can be a health concern unless the condensation is removed from the ducts.

Water-Source, Forced-Air Heat Pumps (Water-to-Air Heat Pumps)

Remember, a heat pump is a refrigeration system in which the cycle can be reversed. In other words, it is an air conditioner that can reverse its refrigerant flow to take heat either from the space it's serving (the load) or from the outdoors or source medium.

FIGURE 3-2 Apartments in Millennium Tower have geothermal heat pumps in vertical units (right) and console units (left). (*Photo by ClimateMaster*)

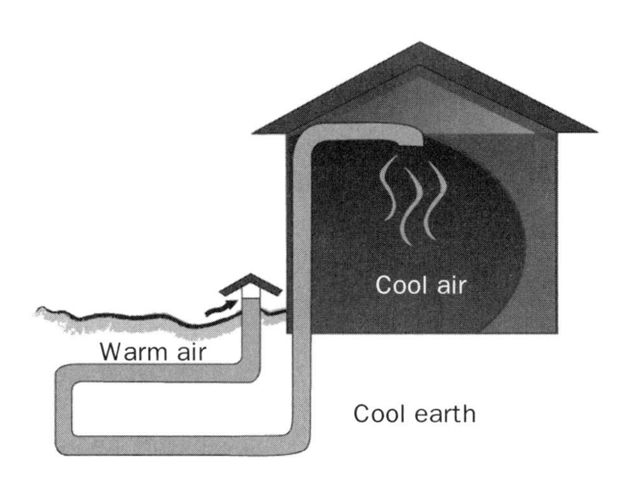

FIGURE 3-3 A series of large pipes or ducts can be run underground to form a simple forced air geothermal system. These are normally considered do-it-yourself projects but they have limited applications. (*Sarah Cheney/Egg Systems*)

Water-source, forced-air heat pumps are by far the most popular and widely used heat pumps at this time, as shown in Fig. 3-4*a*. In a retrofit situation, they can normally be placed where the system's air handling unit (indoor section) is currently located, as in Fig. 3-4*b*. A garage, an attic (or above a ceiling), a closet, and a mechanical room are all suitable.

There are two main differences to prepare for: (1) The physical size and weight may be larger than conventional systems, and (2) the air typically flows through the side and out the end, unlike a standard air handler (see Fig. 3-5). Both of these issues are usually overcome easily during the initial planning stage of the project.

Remember that it's a good idea to cover exactly how and what will be different right up front. A good conversation between the home or building owner and the contractor, including some drawings, will go a long way to prevent tensions later.

Outside, the original condenser is removed from the side of the building. This will often leave an electrical disconnect switch that seems to have no use (see Fig. 3-6). If you have chosen a pump and reinjection system, this power may be used for the pump. If you have no use for the outside power at this time, you will likely find some use in the future, such as an electric car!

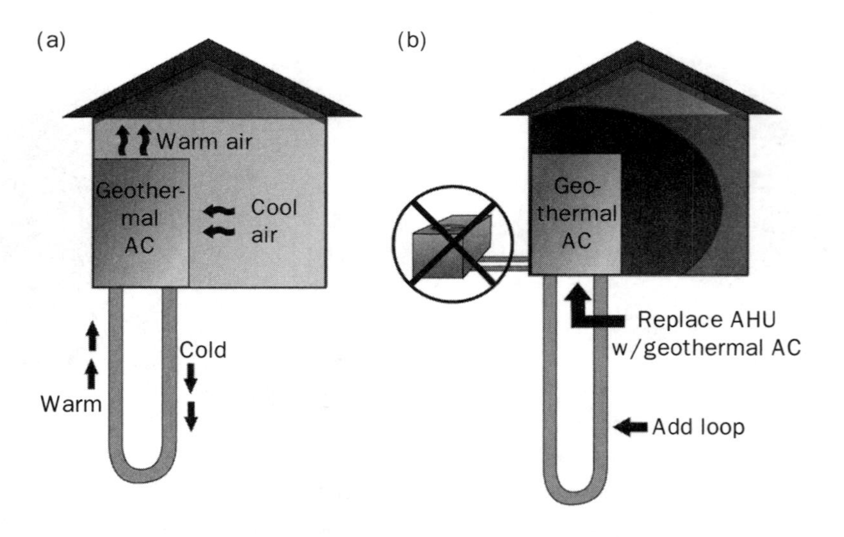

FIGURE 3-4 In order to retrofit a standard split air-conditioning system, (a) the outside unit is removed, and (b) the inside unit or air handler is replaced with a geothermal air conditioner that is connected to the ground loop or well. (*Sarah Cheney/Egg Systems*)

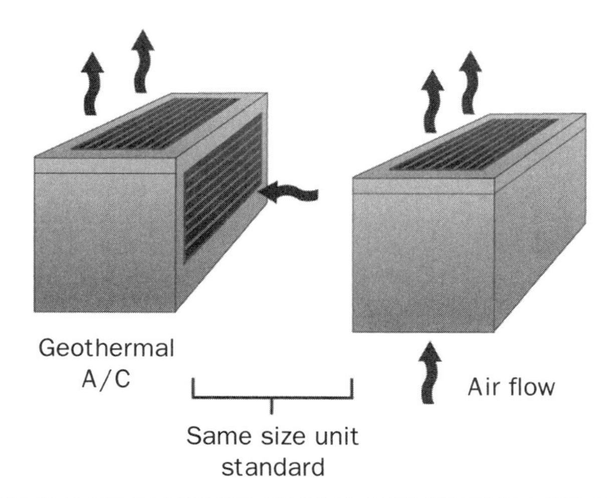

Figure 3-5 Generally, a geothermal air conditioner is about the same size as, or just a little larger than, a standard air handler of similar cooling or heating capacity. (*Cheney/Egg Systems*)

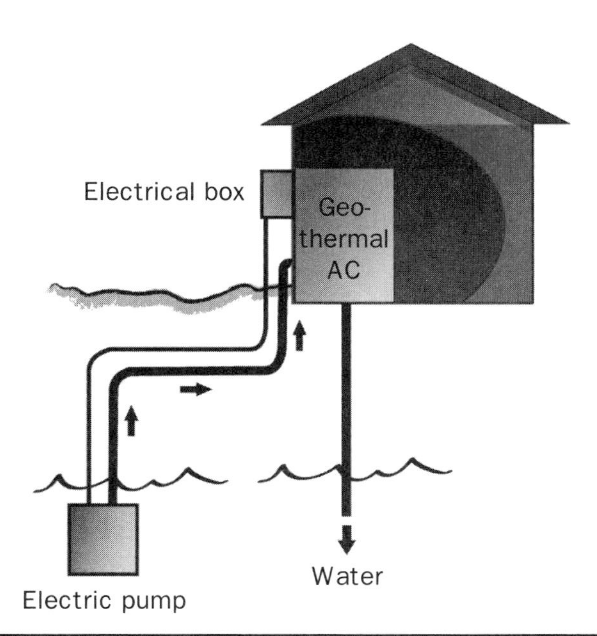

Figure 3-6 In many applications, the power previously used for the outside condenser can be routed to run the pump in a well-type open-loop system. (*Sarah Cheney/Egg Systems*)

Direct Expansion Geothermal Heat Pumps

Direct expansion (DX) geothermal systems are gaining in popularity throughout the industry and the world, in large part because of their high efficiency, ease of installation, and low maintenance. They work under the same principle as the water-to-air heat pumps, but instead of using water as a primary loop for heat transfer, these systems use copper lines installed in the earth and filled with refrigerant, such as R-410A (see Fig. 3-7). R-410A does not damage the ozone layer, unlike many past refrigerants, although it does have the potential to be a greenhouse gas that is much more potent than carbon dioxide.

DX systems require less excavation to install. Theoretically, speaking from a thermodynamic perspective, they are more efficient because one step of the heat exchange process is removed. In fact, according to Monroe, New York-based installer Total Green, DX systems are up to 30% more efficient than other competing earth-coupled technologies, and they readily achieve **4.5–5.0** coefficient of performance **(COP) for heating** and up to **33 SEER for cooling.** However, test data at this time

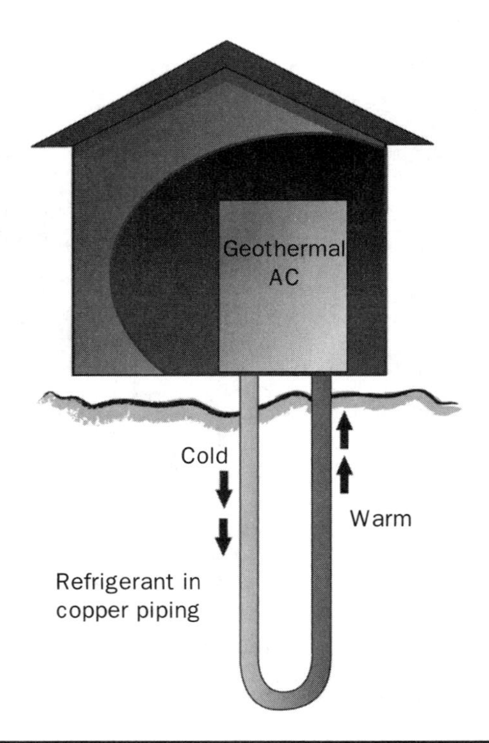

FIGURE 3-7 A direct expansion geothermal system uses refrigerant-filled copper lines directly in the earth. (*Sarah Cheney/Egg Systems*)

cannot definitively prove that any one system is best, and certainly not for all applications and locations. Often, a combination may work best. There are a few more items to consider with DX geothermal systems.

In areas with acidic soils, the copper pipes in DX systems can be at increased risk of corrosion. Some installers will therefore recommend against that type of plan, although it is possible to shield the pipes.

A heat pump uses more refrigerant in certain phases of operation than others. This is normally remedied by using a receiver or tank that is integrated into the system to hold the excess refrigerant. Since DX geothermal systems use a considerable amount of copper tubing in the earth to gain their high efficiency rating, there is a much larger refrigerant receiver than standard systems. This is not a problem unless the systems develop a refrigerant leak. The good news is that R-410A refrigerant is relatively benign. If accidentally released into the ground, there should be no lasting persistent environmental effects since the refrigerant evaporates, according to the material safety data sheet (Fig. 3-8). It is currently about 25% more expensive than R-22 refrigerant, which has been dominant for a number of years, although it is being phased out due to its impact on ozone. As with any system, good installation and service practices should be followed to minimize leaks.

Section 6, excerpted from DuPont™ Suva® 410A Refrigerant SDS. This SDS adheres to the standards and regulatory requirements of Great Britain and may not meet the regulatory requirements in other countries. Before product use, read entire DuPont safety information.

SAFETY DATA SHEET according to Regulation (EC) No. 1907/2006

DUPONT

DuPont™ SUVA® 410A Refrigerant

Version 2.0
Revision Date 10.08.2010 Ref.130000000570

This SDS adheres to the standards and regulatory requirements of Great Britain and may not meet the regulatory requirements in other countries.

6. ACCIDENTAL RELEASE MEASURES

Personal precautions	:	Evacuate personnel to safe areas. Ventilate the area. Refer to protective measures listed in sections 7 and 8.
Environmental precautions	:	Should not be released into the environment.
Methods for cleaning up	:	Evaporates.

The information provided in this Safety Data Sheet is correct to the best of our knowledge, information and belief at the date of its publication. The information given is designed only as a guidance for safe handling, use, processing, storage, transportation, disposal and release and is not to be considered a warranty or quality specification. The above information relates only to the specific material(s) designated herein and may not be valid for such material(s) used in combination with any other materials or in any process or if the material is altered or processed, unless specified in the text.

FIGURE 3-8 The refrigerant R-410A is used in most contemporary HVAC systems. Note that the Material Safety Data Sheet lists the cleanup method as evaporation. (*DuPont*)

Water-to-Water Heat Pumps (Heat Pump Chillers and Boilers)

Water-to-water heat pumps are among the most versatile of the geothermal product lines. However, this technology was not eligible for tax credits until 2010. When the stimulus package was introduced, ClimateMaster and others lobbied lawmakers, explaining to them that this technology can be used for much more than pool heating.

To understand the benefits of water-to-water heat pumps, we need to understand the restrictions associated with refrigerant or DX systems. Remember that DX units are the systems wherein the coils that are serving the load (e.g., cooling or heating air for a building) have refrigerant inside of the tubing performing the heat transfer; thus, the refrigerant is expanding, evaporating, or condensing right inside the very coil that is in the air stream. You can see where the term "direct expansion" comes from.

In a water-to-water device, as in Fig. 3-9, the refrigerant systems are usually inside of the equipment and the heat is transferred to a liquid, such as water, glycol, or brine, before exiting the equipment to go to its final load. Though this type of system is called a *water-to-water heat pump*, meaning it can both chill and heat the liquid, it is also appropriate to call it a *chiller* or a *boiler*. The terms *chiller* and *boiler* imply that the piece of equipment is performing the job of either

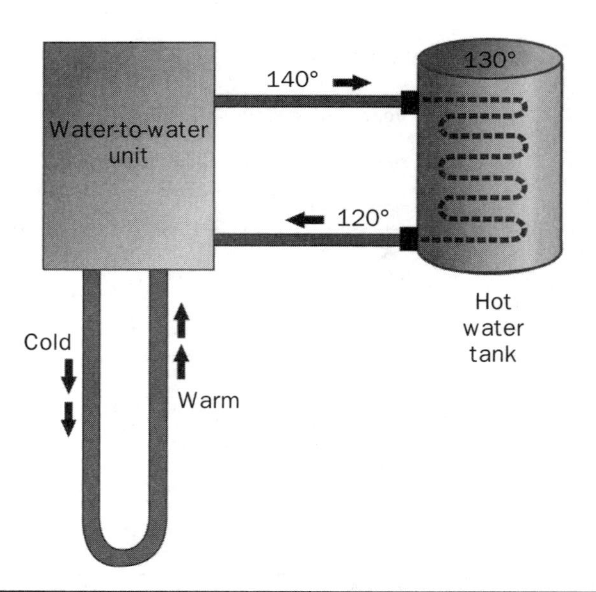

FIGURE 3-9 Water-to-water geothermal heat pumps can be used to heat or cool almost anything. (*Cheney/Egg Systems*)

chilling or heating a liquid medium for delivery to an intermediate or to a final load.

The restrictions that are imposed upon DX systems are a result of the delicate balance of what is called "phase change." This is the marvelous act of changing from a liquid to a vapor and back. This is why refrigeration works. At the risk of becoming too technical for the intent of this book, we will share an example for you to consider: It takes 1 Btu to raise a pound of water from 211°F to 212°F at sea level. It takes 970 Btu to make the jump to 213 degrees and produce steam. There is 970 times more energy in a pound of steam under these conditions than in a pound of water. We call that the latent heat of vaporization and it's what gives refrigerant its ability to carry large amounts of heat energy.

In many situations, if the parameters change too much in the load, the components of the refrigeration system become overloaded and will not, for example, evaporate the refrigerant fast enough. When that happens, the compressor may fill with liquid and be destroyed (liquids do not compress). Suffice it to say, when multiple load DX systems are properly engineered, they can perform safely and effectively.

Chilled water systems (typically used in large commercial structures) have the advantage of no phase change (the change from liquid to vapor or vice versa). They simply typically provide chilled liquid at 42 degrees (example of an air-conditioning load) to as many fan coils as the designer desires. (A fan coil is an air-moving device normally connected to ductwork that uses chilled water as the heat transfer medium.) The water takes on heat and becomes useless for heat transfer at some point above 50 degrees, depending on conditions. Often times, these systems will have an additional heat or reheat coil with separate piping from a separate piece of equipment supplying the heated fluid. This is because in precise temperature and humidity control situations, the system must be able to reheat chilled air for humidity removal and control. We won't be going into how that works in this book.

One of my Egg Company divisions is currently engineering a residence for this type of system. The home has two 70-ton water-to-water heat pumps. One provides chilled water to only 27 different fan coils in the residence. The other supplies heated water (135 degrees) to the fan coils for the job of reheating air, and the remainder of its capacity goes to the job of domestic water heating needs. These two systems are in load-sharing harmony as the boiler uses the waste heat from the chiller, returning the primary coolant or ground-source loop back to its original temperature (see Chap. 5 for more on load sharing). The loop coolant's final stop on its circuit takes it through the 40-ton water-to-water pool heat pump exchanger, further chilling the water as more heat is extracted to heat the pool (more on this later in this chapter).

The versatility of this system is such that it is preferred in larger commercial installations. The initial cost is usually quite a bit more than DX systems. This is due to more expensive piping, insulation, valves, controls, and equipment, as well as increased labor costs.

Applications of Geothermal Heat Pumps

Pool Heaters

The most impressive savings for any geothermal applications, in my experience, have been seen in geothermal or ground-coupled pool heating, which can readily save 75% on energy bills versus traditional gas systems. This is significant because it is not uncommon to find a commercial pool that is spending $50,000 a year on natural gas or propane. After installing a geothermal heat pump, they will see the gas bill eliminated (no more need for gas!), replaced by an electric bill of between $5000 and $10,000 a year. Properly engineered, a geothermal pool heat pump can provide all the heat the pool will need to maintain a comfortable swimming temperature year-round.

For geothermal pool systems, an open loop (pump and reinject) should be carefully considered if possible. If the thermal extraction from the soil around the loop field is not recharged in a closed-loop application, the system may lose its capability to heat. This is where proper training and engineering are imperative.

IMPORTANT NOTE: *Make certain your contractor has the credentials to do your system.*

Recently, I was asked to visit a geothermal pool heat pump manufacturer's facility to talk about several topics, from distribution to engineering. Among other things, the general manager commented about the serious concerns they have with the proper implementation of the technology, with so many jumping into geothermal at this time. As we conversed for much of the afternoon, I shared with him some of the critical guidelines that we use in our installations. Among these was the necessity of proper pressurization of the groundwater exchanger at all times, to prevent fouling. He responded, in a self-deprecating way, that they had designed the valve system to keep the poolside titanium exchanger pressurized at all times, because it substantially improved the system's life span. He wondered why they had not considered doing that on the geothermal side before our conversation, and said they intended to add that to their specification guidelines for their equipment. It seems we are always learning from one another.

Domestic Hot Water

Domestic hot water is one of my favorite geothermal technologies. Recently, I had a new geothermal customer call me from Melbourne, Florida. He has a natural gas-fired hot water tank that has been tied into his new ClimateMaster geothermal air conditioner. The heat pump has an integrated hot water generator that provides a sizable portion of his hot water needs. He called to ask if, since it was January, he should turn off the hot water generator and let the natural gas take the lead? His thinking was that the energy expended in heating the water from the geothermal system's compressor would be more effectively put to use heating his home. He thought the natural gas was more efficient, dollar for dollar, than having the geothermal heat pump work a little harder to produce that extra heat. He was certainly surprised when I told him that the geothermal heat pump did work hard, but that it did its work for 80% less money than the gas.

People will spend $5000 to install a good *solar water heating system* on their home (although tax credits can help with that). There is really nothing more efficient for heating hot water that we are aware of (see Fig. 3-10). I have a great solar hot water collector on my home

FIGURE 3-10 Solar thermal panels can efficiently generate hot water for household needs, with a much quicker payback period than photovoltaic solar panels that create electricity. (*Photo by Brian Clark Howard*)

right now, installed only two years ago. I live in the Sunshine State, and my solar system works great even when it's partly cloudy and freezing outside. As I write this section, I'm sitting in my van, working on my laptop while my son's Boy Scout Troop is running around in drizzle and 33-degree temperatures, playing manhunt and trying to stay warm.

When it's completely overcast or nighttime, the solar thermal heater will not be able to generate hot water. When freezing is a threat, the freeze protection feature in my system will engage and take heated water from the tank and circulate it through the collector to prevent bursting of the copper pipes. This makes my collector essentially an outdoor heater (see Fig. 3-11). That means that I am paying for electric heat, which has a COP of less than one. This is bad because the electrical backup heater within my tank has to heat water to keep it from freezing. That uses energy, but for us here in Florida, freezing is rare, and so the increased electrical consumption may not even be felt. In freezing climates, the collector would likely need to be winterized, or in other words, drained of water.

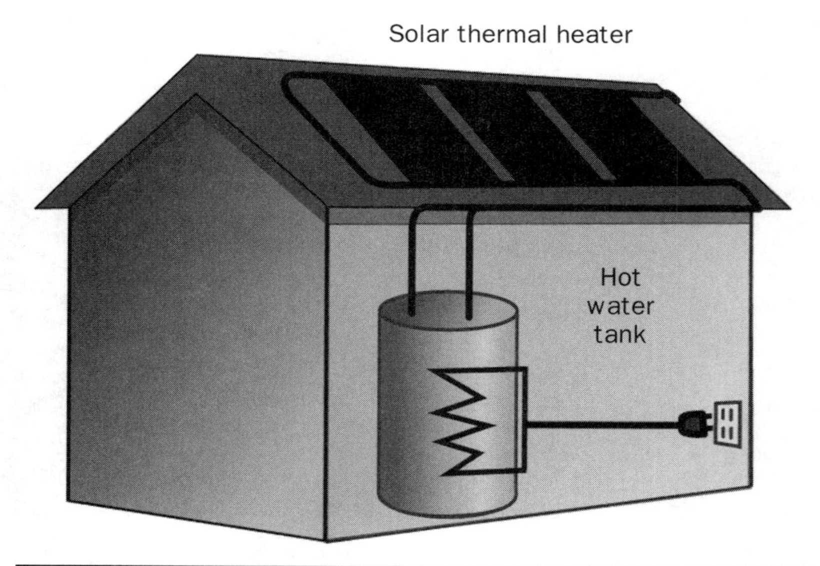

FIGURE 3-11 Solar thermal heating is very effective and energy efficient. The tank usually has a backup heating source, such as an electric heating element or gas burner. (*Egg Systems*)

As a result, I am installing a dedicated geothermal hot water heat pump, just for my 80-gallon tank. The water-to-water heat pump is only about the size of an apple box (see Fig. 3-12). And the energy that my heating element is using now will power the heat pump four times over. Since I already have a geothermal air-conditioning system, all I am doing is branching over the piping from the system and load sharing with the rest of the house. In all, this will put my four largest energy drains on one load-sharing loop. And no more guilty conscience about late night showers! (You know, no sun, no hot water, expensive gas or electric heat.)

Process Cooling and Heating

There are many opportunities for process cooling in all facets of industry. A couple of examples include plasma cutters and extrusion presses. Such hard-working equipment is often air cooled because of ease of installation. But, like standard chiller and boiler heat pumps explained above, these could be hooked up to earth-coupled systems to achieve impressive energy savings and take advantage of available tax credits, as in Fig. 3-13.

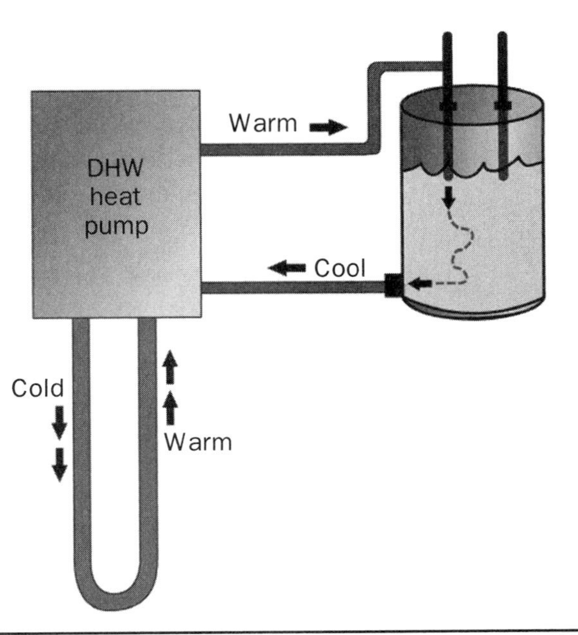

FIGURES 3-12 A dedicated geothermal hot water heat pump can provide all of the hot water needs for your home or business day or night, any season of the year for about 25% of the cost of gas or electric heat. (*Sarah Cheney/Egg Systems*)

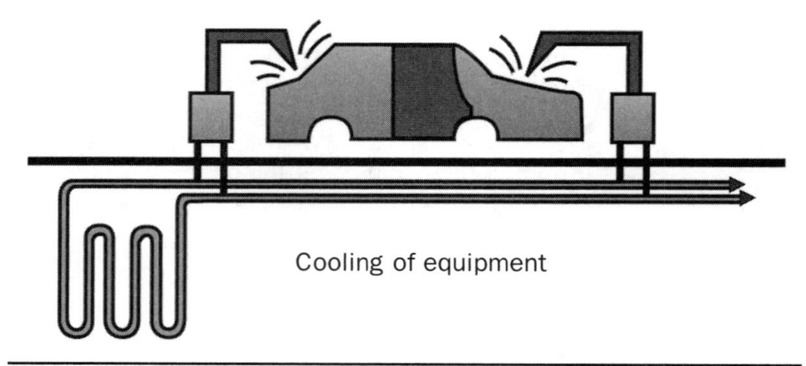

Figure 3-13 Most manufacturing processes need cooling of some sort for the equipment. Geothermal cooling is becoming increasingly popular for these applications. (*Sarah Cheney/Egg Systems*)

Rooftop Equipment

There are thousands of shopping centers and commercial buildings with rooftop package equipment already installed. These are favorable for the ease of installation and removal. A crane can pick up an old unit and deliver a new unit in less than half an hour. When a unit is in need of service or maintenance, it can be performed without any inconvenience to the patrons below.

In this case, upgrading to a geothermal system involves a simple process of installing the necessary piping across the top of the roof, in close proximity to the main equipment, as in Fig. 3-14. Then, the equipment can be upgraded to geothermal rooftop equipment, either as the need arises, or all at once.

Modular or Piggyback Units

Conventional *modular* or *piggyback* HVAC units are typically used in high school portables, construction trailers, and cellular tower equipment rooms (see Fig. 3-15*a*). These are among the least efficient air conditioners due to the limited size of the outdoor coil by design. Therefore, they can be good candidates for replacement with earth-coupled technologies.

In situations involving multiple modular units, as in a school (see Fig. 3-15*b*), the geothermal supply and return piping can be run underground before setting the portables, providing enough stub ups

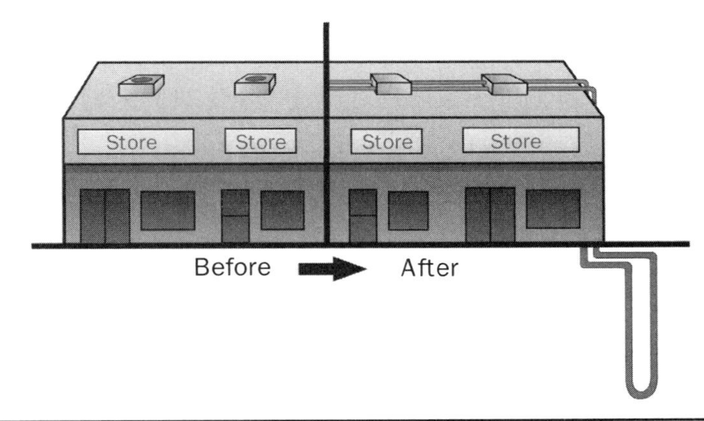

FIGURE 3-14 When a commercial property owner decides to upgrade to geothermal cooling and heating, he or she can install a ground loop or well large enough for the entire property and convert the spaces over to geothermal air-conditioning by attrition, connecting each upgraded unit to the geothermal system as needed. (*Sarah Cheney/Egg Systems*)

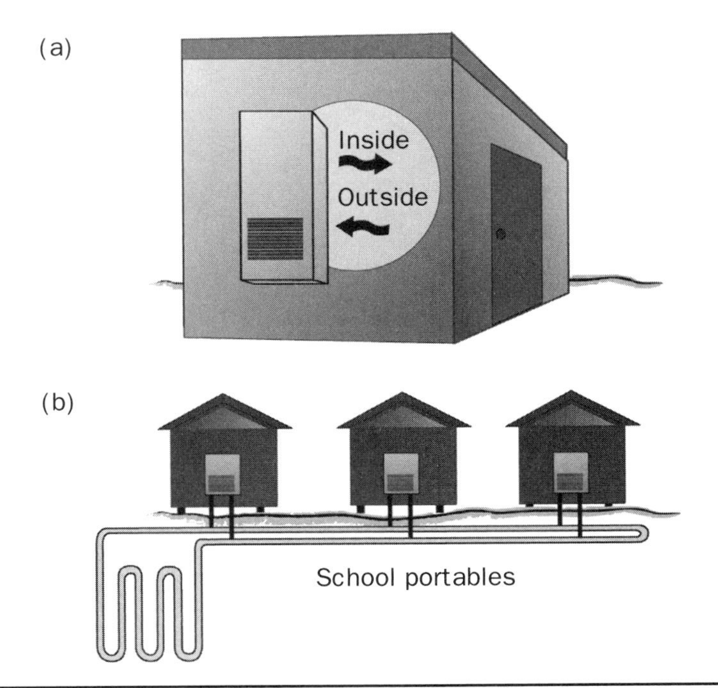

FIGURES 3-15 Modular air conditioners are often found on (*a*) the sides of cell towers, equipment buildings, or (*b*) school portables. Such applications are suitable for a geothermal upgrade. (*Sarah Cheney/Egg Systems*)

to accommodate future needs. In situations such as a cell tower, the cooling hours are often significantly higher due to internal loads, which means the payback time can be surprisingly short.

Package Terminal Heat Pumps

Packaged terminal air conditioners, or PTACs, are the bulky box units that are found in most hotel rooms, as well as numerous other applications. These are the closest systems, commercially speaking, that you'll find to a common grade window unit. These systems have historically offered low efficiency due to the limited size of the unit, and once again the outdoor coil (see Fig. 3-16). In the case of geothermal replacements, piping is run to the vicinity of each unit, and each unit typically has its own circulating pump.

Vertical Stack Modular Units

In multistory apartment buildings and hotels, vertical stack modular units give a high-end appearance and functionality. They are easily roughed into place and become part of the piping circuit, as in Fig. 3-17. Vertical stack modular units can help a building achieve a higher LEED rating. All hotels require fresh air, which is easily supplied through the PTAC unit; when upgrading to a modular unit, a fresh air duct will need to be added to the system.

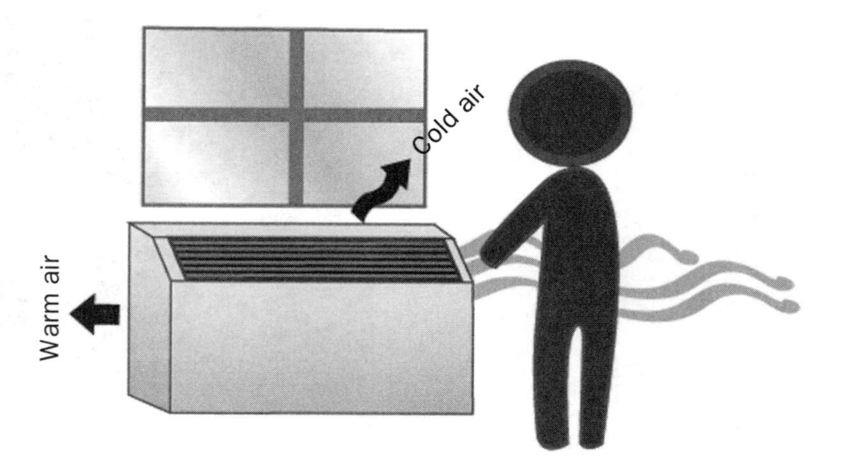

Figure 3-16 Hotel room air conditioners are typically relatively inefficient. PTACs, as they are called, can be upgraded or specified as new equipment in these applications. (*Sarah Cheney/Egg Systems*)

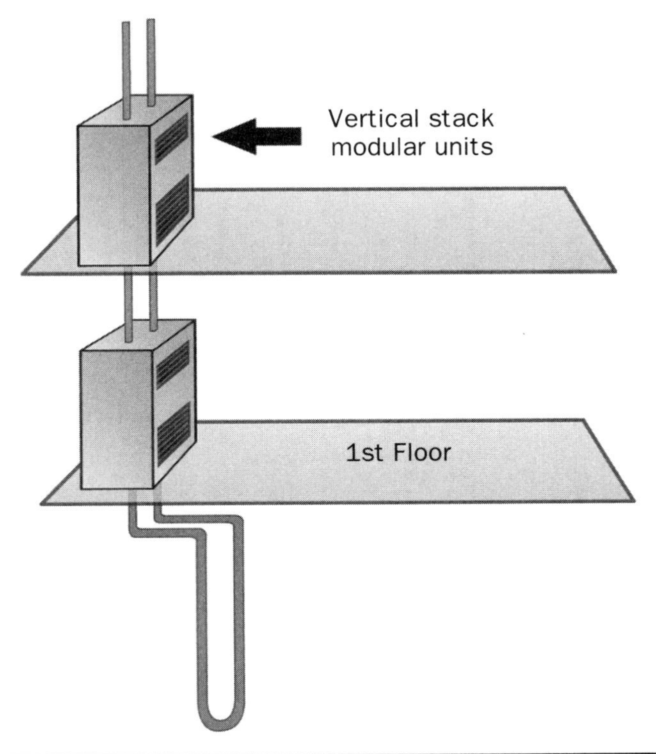

FIGURE 3-17 In high-end hotels and office buildings, vertical stack modular heat pumps become part of the piping circuit. (*Sarah Cheney/Egg Systems*)

100% Fresh Air Equipment

You guessed it: 100% fresh air-conditioning equipment does what the name implies (see Fig. 3-18). This can be accomplished through several methods. Geothermal-sourced 100% fresh air equipment has a significant energy advantage over other air-sourced or cooling-tower-sourced equipment. This equipment must typically run during the occupied hours of the building it is servicing, which can be 24 hours a day in many situations, such as hotels and hospitals. When the climate has high heat and humidity this can be the single largest electrical load in a building, so the energy savings by going ground sourced are dramatic.

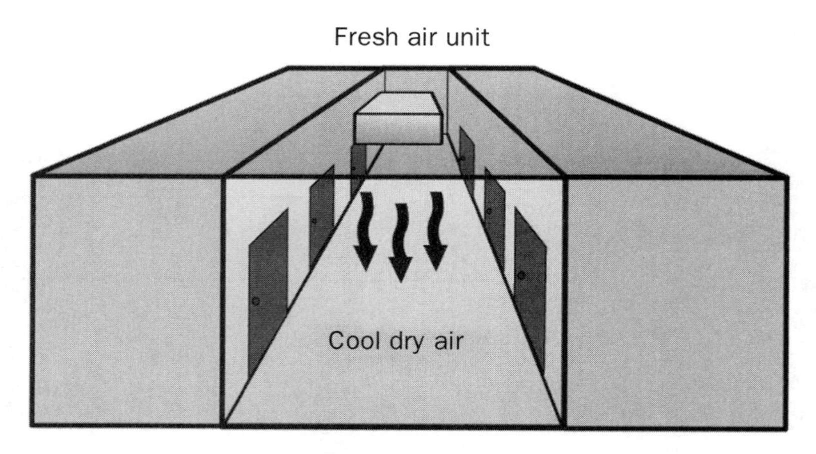

FIGURE 3-18 100% fresh air equipment has a high energy load because the air conditioner is treating the outside air rather than recirculating conditioned inside air. Geothermal can greatly reduce the energy consumption of 100% fresh air equipment. (*Sarah Cheney/Egg Systems*)

Refrigeration Systems

Almost every significant refrigeration system in commercial (and some residential) applications has a water-cooled option offered by the manufacturer. Some examples include ice machines, walk-in coolers, ice cream machines, vending machines, and cryogenic systems. Water-cooled models often cost less because they need less metal for the water-cooled exchangers, due to the higher efficiency of water in heat transfer. These water-cooled devices can be used with geothermal systems.

What does the future hold? Many more appliances may become part of the geothermal load-share circuit. Water-cooled refrigerators or freezers would operate more efficiently and quietly, and life expectancy would increase while maintenance and repair costs would decrease.

Hybrid Systems

Hybrid systems are a good idea in some commercial situations where an increase in efficiency can be achieved through the use of a supplemental geothermal source. Since most air-conditioning systems are ultimately air cooled, they are limited in efficiency to the amount of air that can be rejected to the outside. If it's a warm day, air-cooled equipment can be inefficient. For example, if the temperature outside is 95 degrees, your system will be unable to get the condenser refrigerant line below that temperature. The earth's temperature, however, will usually be much lower. In a hybrid system, an exchanger is placed strategically to produce a cooler condenser temperature, and therefore higher efficiency and capacity.

In many climates that may be classified as heating dominant, such as Canada or Norway, the geothermal portion is often sized to handle about 60–80% of the heating load, with the remaining peak load handled by electrical resistance heat strips or gas burners. In this way, the geothermal ground source can become more standardized and can offset most of a 6000-hour heating season. The reason the geothermal loop is not sized for 100% of the heating load has to do with economics, or what we often call the point of diminishing returns. At some point, the loop size might have to be doubled to get the last 20% of seasonal heating, and this just may not be cost-effective.

In cooling of large commercial structures, there are circumstances too varied to name here that may justify using an air-source technology seasonally, periodically, or for redundancy or peak capacity.

Superefficient DC HVAC

OK, we have geothermal heat pumps that now achieve energy efficiency ratio (EER) ratings of 30. Can it really go higher? I'm glad you asked because I have another story about that...

My dad is a bit of a techno geek, which explains some of my problems. ...Anyhow, he had an awesome electronics shop full of gadgets, motors, and wires. Among these was a powerful alternating current-to-direct-current rectifier about the size of a 20-inch tube TV. This rectifier had a myriad of potentiometers, dials, and switches. What that means to a dangerously curious boy of 10 is hiding pets and electronically powered devices. (For those of us who love animals, I never actually did anything cruel or mean, and they were never harmed.)

I did, however, figure a lot out about the limitations of alternating current. To make it simple, alternating current voltages go from the rated voltage of say 220 volts (V), down to zero and back up again 50–60 times each second. If that sounds confusing, listen to this: When the voltage gets to zero, it actually goes to negative 220, then swings all the way back up again to the positive side. It actually works really well, especially when we talk about the ease with which voltages may be changed in power distribution to reduce losses due to heat and other factors.

Batteries, as well as photovoltaic and fuel cells, produce direct current. It's rather easy to vary the voltage on a direct current device, and more difficult to effectively do it on an alternating current device. On alternating current, we typically vary the voltage frequency but not the voltage itself.

There is a type of motor called a *universal motor* or an *AC/DC* motor that will run on AC or DC (usually the DC voltage required is lower than the AC). Portable power tools, vacuum cleaners, and sewing machines contain these motors, although they often aren't intended to run directly on DC. (Mom, I still maintain that I don't know where that sewing machine motor went.) They are popular for tool applications because universal motors also happen to be very powerful for their size and weight. A typical 115-V universal motor will run just as well on about 30 V DC.

I was fascinated with the capability these motors had when used with direct current. I found that I could drop the speed down to 25% or less, and run it right up to nearly twice the speed that it was able to attain on alternating current. I do have a point . . .

Now that childhood tinkerers like me are approaching 50 years of age, you're going to end up seeing air conditioners use this same technology. They'll have quieter, variable speed compressors and fan motors with all the bells and whistles. Imagine having only two residential sizes from which to choose. You could buy a nominal 5-ton heat pump that can operate in a range from 2.5 tons to nearly 7 tons. Won't that be nice for entertaining? A couple of extra tons of cooling or heating capacity for those occasions when you really need it. And if you're like me, I've always disliked even a one- or two-degree temperature variation between on and off cycle. With DC air-conditioning, the unit runs most of the time at exactly the speed it needs to go to keep you at exactly the temperature you desire. And the great part is the efficiency: I think we're going to be in the 40 EER range or better for a seasonal average.

Keeping the Cows Cool

How about livestock? Have you ever considered the comfort of dairy cows? What about pigs? I would have never guessed this, but this is quite a problem. There are facilities spending hundreds of thousands of dollars a month to keep their livestock cool in order that their milk production, growth rates, and birth volumes will be maximized. The businesses with which I am involved are in tests with universities, manufacturers, and power companies to offset these high costs with geothermal earth-coupled technologies.

What does that mean to you and I? I would sure enjoy a reduction in cost for a gallon of milk the next time I go to the grocery store. That's really what it's about: reducing our energy costs so we all can keep more of our hard-earned money.

Summary

As we have seen, geothermal earth-coupled systems are adaptable and come in various flavors, from simple passive systems that are accessible to the DIY crowd to more complex water-source, forced-air heat pumps; DX heat pumps; and water-to-water heat pumps. The technology works well for small applications, or can be scaled up for large commercial or industrial needs.[1] In addition to being a quiet, low-maintenance solution for heating and cooling, geothermal systems can provide domestic hot water, heat pools, cool off industrial equipment, and more.

[1] In the case of large-scale heating or cooling needs, the geothermal heat pumps may require, or at least work better with, three-phase electricity. Such service is widely available and is already the standard for big conventional HVAC systems and industrial-grade motors.

CHAPTER 4

Earth Coupling through Ground Loops

He who builds to every man's advice will have a crooked house.

—Danish Proverb

I recently completed an addition to my house in Florida. Many of the energy-saving measures that went into this construction are mentioned throughout this book. One bit of technology I used is occupancy sensor light switches for the rest rooms, showers, laundry room, closets, and vanities. I was very pleased with myself the day I walked through the addition and noted how the lights came on as I entered each room, and went off a few seconds after I exited. Of course, the bedrooms had standard switches, as I realized that there would be times when there would be little motion, but the children would need the lights to stay on, such as during study.

Later that week, as I took the opportunity to use one of the rest rooms, I noted that the lights went out only a few moments after getting comfortable. I cheerfully waved a hand in the air to activate the sensor and bring my light back on. But then it went off again. A little less patient, I waved my hand again, determined to keep my hands and head moving a little this time. Much to my chagrin, I wasn't dancing quite enough for the sensors' applause meter, or whatever you want to call it. By the time the whole ordeal was finished, I was bobbing and weaving my head and arms from left to right, and up and down, all the while making up the words to some ABBA song from the 70s that I vaguely remembered, pretending like I was OK with this.

It's two years later, and I'm finally replacing the motion sensors with new units, which have adjustable "on" times. Ours are set for five minutes. Oh, I forgot to mention, my kids just switched the things to "on" and never shut them off again, so the whole energy-savings thing backfired on the lighting. The moral of the story is the reason for this chapter: Please take the time to ask questions and make sure you and

FIGURE 4-1 Installing a ground loop for a geothermal HVAC system. (*Photo by Egg Systems*)

your installer are on the same page about your needs, expectations, and lifestyle. Allow me and other professionals to tell you what works, and what doesn't, backed up with years of experience and data. I wish my electrician would have said, "Don't use that switch there; you'll end up learning to dance in the dark." That being said, educated consumers and educated contractors are the best defense against future problems.

Getting the Load and Loop Size Right

Armed with this true parable, note that this is perhaps the most important chapter of this book. With 20 solid years of installation experience of geothermal systems, and having seen many of the mistakes that can be made with this technology, trust me when I suggest that you take the time to ensure you have an understanding of your specific situation.

When trying to pair up a particular heating or cooling load to a ground-coupled design, there are numerous considerations that are imperative to be taken into account.

Important Note: *Thermal conductivity of the soil may need to be checked, using a test bore, and load hours may need to be determined. Please note that I said "may" need to be determined. There are also many established parameters that I will share in this and other chapters.*

An example of an error that is commonly made involves the miscalculation of the load hours compared with the loop capacity. Imagine a contractor who has begun to become successful in residential loop installations. Now, this contractor is asked to install a commercial geothermal system for cooling the vacuum chambers in a laser crystal coatings facility. Having sized many loops by interpolation based on instantaneous load capacity, like most other types of refrigeration, the contractor selects a 40-ton loop, figuring on a 100% safety margin beyond a base of 20 tons, and assuming this will more than suffice. In truth, proper load calculations would show that the loop needs to be designed for closer to 400% of a typical 20-ton system because it runs two to three times more hours per day than a standard commercial load. In this case, that extended load is a result of multiple shifts per day.

The loop mentioned above, having been grossly undersized, would begin to start showing signs of failure inside of the first month of operation, no matter if it were summer or winter. Service calls would be placed as the loop temperatures increased to 105°F and above. The loop technician would perform a leaching of the ground loop, which involves replacing the water in the loop with cool water for days at a time, hoping to cool down the ground surrounding the loop. This is counterproductive for more than one reason, including the wasting of water, diminished efficiency, diminished capacity, and a loss of labor hours. Unfortunately, there is no easy answer for this situation at this point. The contractor will lose, the business owner will lose production time, and ultimately, the geothermal or ground-coupled industry will lose credibility.

The same thing can happen in a residence if all the right questions and angles are not explored. The cooling hours per year of a southern region can be many times the cooling hours of a northern region. And the heating hours of a northern region can be many times that of a southern region. Without the proper training, information, and experience, the design and installation will likely have flaws.

In Utah, the granite earth is more conducive to vertical boring and, with the proper grout, is a good conductor of heat, as is the soil in many areas of the world. You will also find that the cooling and heating hours are similar. In southern Florida, the sandy earth is wet, and saturated sand is a good conductor of heat. But because of low heating hours, a plume of thermal retention builds and begins to store larger and larger amounts of heat until the system stops functioning properly. We have seen systems fail after three years of successful operation, killing the ground cover above due to heat. Conversely, the same loop in the same soil but in a different climate might result in exemplary performance.

Piping Material

When I started in the geothermal heating and cooling industry (early 1990s), we were trained to use high-density polyethylene (HDPE) or polybutylene piping. Polyethylene must be butt fused or mechanically crimped for final connection to threaded fitting. Polybutylene piping was socket fused and was generally considered the less expensive of the two recommended installation materials. For many reasons, including poor dependability, the polybutylene piping was no longer recommended for use after 1995. HDPE is widely in use today and has proven many times over to be a dependable material for earth loops. It is also commonly used in natural gas piping due to its low failure rate.

Some other materials are sometimes used, but few have the longevity or low cost of HDPE (see Fig. 4-2). Examples of these are PVC (which is not popular in environmental circles due to the toxic byproducts created in its manufacture), CPVC (chlorinated polyvinyl chloride), copper, and galvanized piping. Some manufacturers recommend that copper be used in conjunction with a sacrificial anode

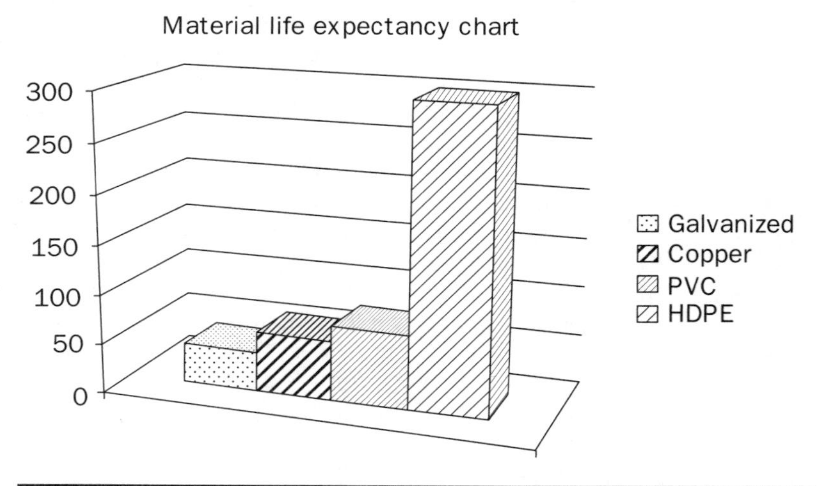

FIGURE 4-2 High-density polyethylene (HDPE) is typically used for geothermal ground loops because of its longevity and relatively good thermal conductivity. (*Egg Systems*)

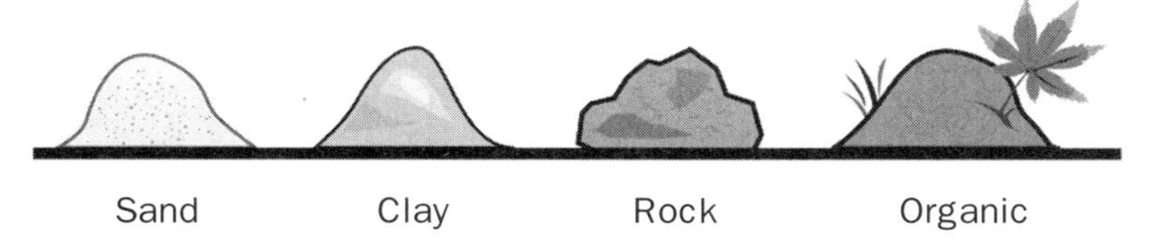

FIGURE 4-3 Different soil types present different challenges and opportunities for installing geothermal systems. (*Sarah Cheney/Egg Systems*)

and/or a thin protective sleeve, such as is used for protection under slab copper. As we stated in the previous chapter, this is the strategy that is often employed for direct expansion systems when the soil is acidic, to prevent corrosion. Still, copper is relatively expensive. No loop material has proven itself in cost and dependability as well as HDPE.

Grout and Backfill

Saturated sand is used as the basic standard for thermal conductivity in geothermal or earth-coupled applications. We've written previously in this book about the conductivity of air versus water (water is 24 times better at conducting heat). For purposes of an analogy, we could compare water to sand and air to silty soil. ...It's also true that silty soil is not good for conduction of heat.

Without spending a lot of time on this subject, suffice it to say that a contractor who does not know what he or she is doing will make mistakes. The wrong soil type, installed the wrong way, will not work properly, even with 10 times the length of piping installed (see Fig. 4-4 for a rough guide). Always use a geothermal engineer to design your system. Help us protect you and this industry.

Manifolds or Header Systems

As with the design of highways, if we don't put enough "lanes" in, we'll end up wasting energy and reducing efficiency. The way to increase efficiency and performance is to make the piping like a multilane freeway with several lanes going the same way, as depicted in Fig. 4-5*a*.

To put this in practical terms, if we have a 10-ton system that requires a thousand feet of 3/4-inch pipe per ton, we can't just put 10,000 feet of pipe in one continuous loop. That would be called a *series loop* and would require too much pumping power. Normally, we would put 10 to 20 parallel loops into the circuit. Each circuit would be between 500 and 1000 feet in length, as in Fig. 4-5*b*. In this way, we are able to take full

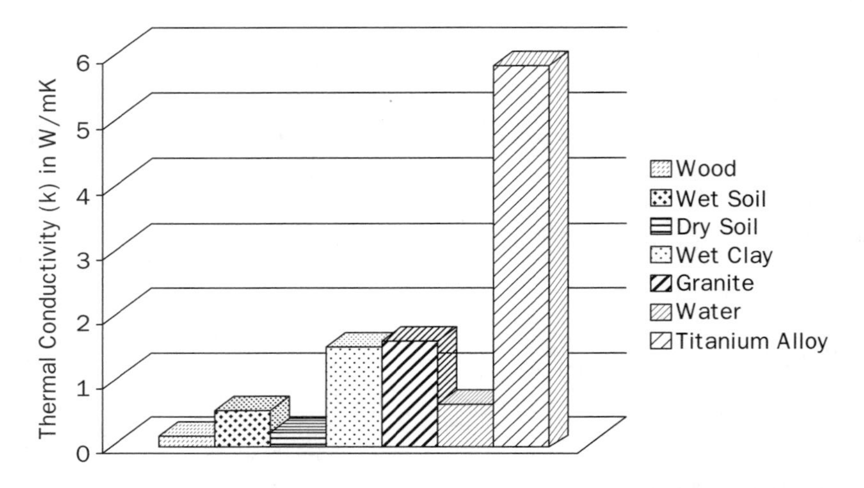

FIGURE 4-4 Thermal conductivity comparison chart. Be sure to know the thermal conductivity of your particular soil. This has a great effect on the length and type of ground loop exchanger that you will have installed. (*Sarah Cheney/Egg Systems*)

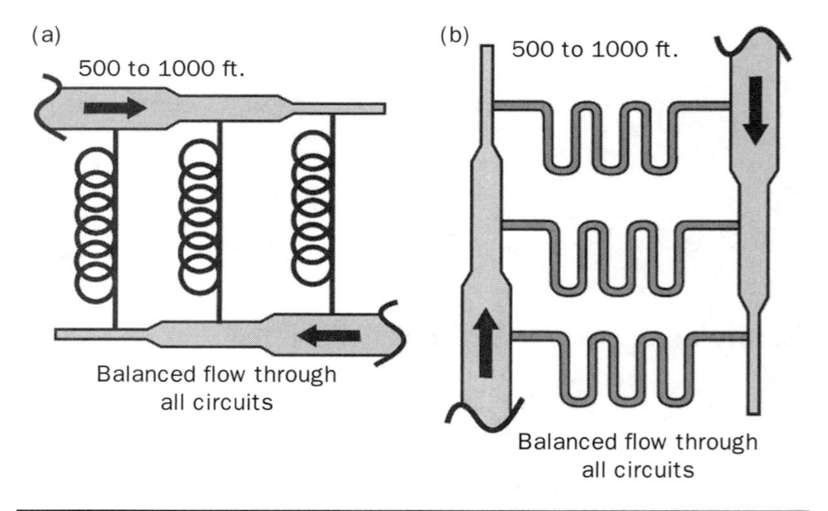

FIGURE 4-5 (*a*) Ground loop exchangers require precise engineering to allow for balanced equal flow through each of the parallel circuits. (*b*) The length of piping per ton can vary by up to 400% depending on many factors. Proper header design is critical to acceptable performance of the system. (*Sarah Cheney/Egg Systems*)

advantage of the 10,000 feet of pipe with only a fraction of the pumping power.

Loop Designs

Figures 4-6*a* and *b* show some of the basic types of ground loops. Certain styles of ground or pond loop require as little as 300 linear feet per ton of heat exchange. Some styles of loop require a thousand or more feet per ton. It's easy to see that a home with 5 or 10 tons of air-conditioning can have a mile or two of piping under the ground.

The problem with this much pipe is that pumping power increases at a ratio higher than the benefit after a certain distance of linear feet. After about a thousand linear feet, a contractor will normally terminate

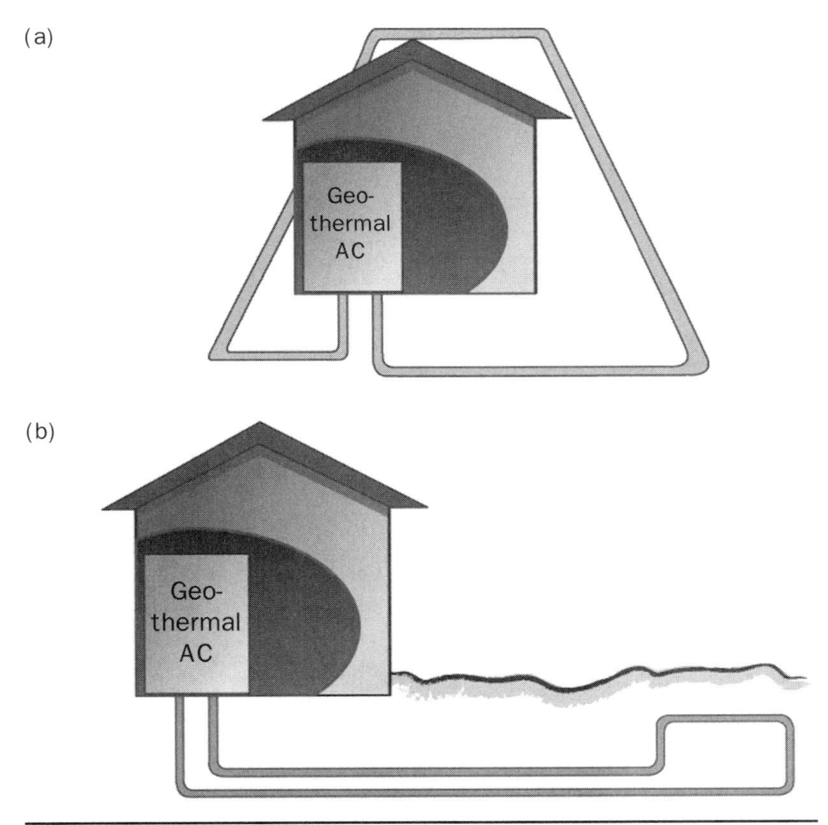

Figure 4-6 (*a* and *b*) Depending upon the characteristics of your property, your loop design can be configured in one of many possible varieties of orientations and configurations. (*Sarah Cheney/Egg Systems*)

either end into the supply or return header. These headers are engineered for precise fluid flow under specific conditions. Many systems will have three to ten or more parallel taps into the headers. These must be carefully balanced for proper flow. If piped or balanced poorly, the liquid in the loop will take the path of least resistance and waste the capacity of the parallel loops. *Again, always use a trusted and certified contractor because he or she will identify these issues before they become problems.*

There are several basic types of earth-coupled exchangers or ground loops. These can be installed in many different ways, some of which are recognized by the generally accepted industry authorities of IGSHPA and GeoExchange. Manufacturers also have a substantial say in this process, as they are the ones who inevitably receive the complaints when something does not work well. ClimateMaster has been among those instrumental in the development of this process and has spent considerable resources to ensure the stability of the industry.

FIGURE 4-7 Pictured are loop terminations from a vertical loop installed in Florida under a parking lot. Note the purge ports, which are used to fill and purge the air out of the ground loop. The threaded fittings at the top will be connected to geothermal equipment inside of the building. (*Photo by Egg Systems*)

Vertical Loops

The most common type of closed loop is probably the vertical two-pipe or U-bend system. It typically uses a 3/4-inch HDPE prefabricated into a U-bend, connected by heat fusion and inserted into a borehole. These pipes are close together, often in contact with one another for much of their length. The borehole can be as shallow as 20 or 30 feet, down to a depth of 500 feet or more. The length of pipe required on average for this type of system varies according to thermal conduction and other conditions of the soil and the loads, but is often 500 linear feet per ton of cooling or heating (as you may recall from Chap. 4, a ton of cooling is equal in energy movement to 3.5 kWh of heat).

Depending on soil conditions and local codes, grout should be used in the installation of vertical loop systems, as depicted in Fig. 4-8. Grout enhances the thermal conductivity of the HDPE pipe and seals the hole to prevent migration of groundwater or pollutants from one aquifer to the other. However, note that some areas prohibit uses of certain grouts.

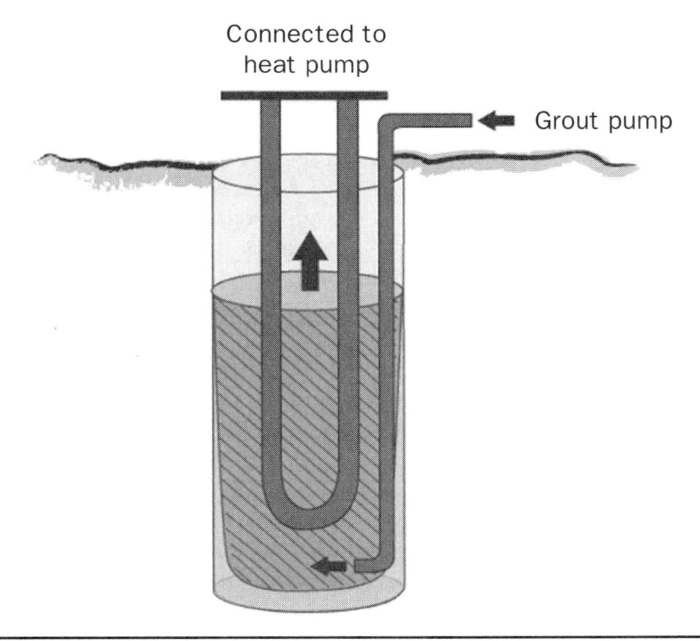

Figure 4-8 To ensure good thermal conductivity, Bentonite grout is injected into the borehole. This fluid mix hardens like cement and provides good thermal exchange between the fluid in the pipe and the surrounding earth. (*Sarah Cheney/ Egg Systems*)

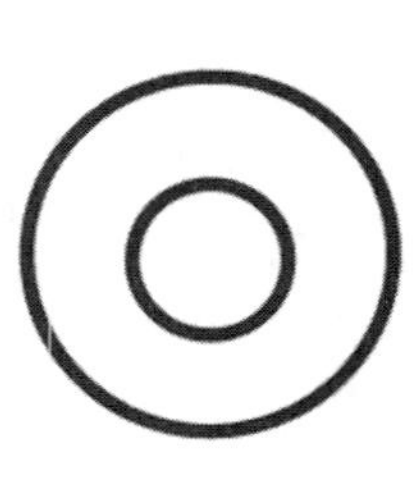

Figure 4-9 There are a number of different styles of vertical exchangers, including coaxial, which is essentially a large pipe around a smaller pipe sealed at the bottom, and complex coaxial, which is less common. (*Sarah Cheney/Egg Systems*)

The primary reasons for the popularity of such vertical piping are the nearly perfect thermal stability of the soil at a depth of 25 or more feet, and the limited area and excavation needed to install the loops. Make no mistake: This is a messy process. But the sheer volume of earth moved is significantly reduced in a vertical loop installation. This system is great for use underneath parking lots in commercial applications, or anywhere space is at a premium. It is also often employed in areas with rocky soil, such as in New England.

A few other types of vertical systems include the *coaxial exchanger* and the *borehole spiral exchanger*. These are not as popular, but can be effective when engineered properly. A coaxial loop is often much like a U-bend system, except the former uses a larger pipe on the outside, and a smaller pipe inserted inside, just several inches short of the depth of the larger pipe. The large pipe is capped at the bottom (see Fig. 4-9).

There are also complex applications that involve several return pipes or channels. This technique is rarely used in the United States, but is sometimes employed internationally.

Horizontal Loops

The most common type of horizontal loop today is likely the *slinky-style loop*, as shown in Fig. 4-10. Given the name for its spiral, slinky-like design, this system was devised in the 1990s from a simple template made from some plywood and 2 × 4s. Put together just so, and properly spaced, it can typically fit a thousand linear feet of 3/4-inch pipe into a trench that is about 80 feet long and 3 feet wide (Fig. 4-11). Often, the earth-loop contractor will excavate a significant area completely, rather

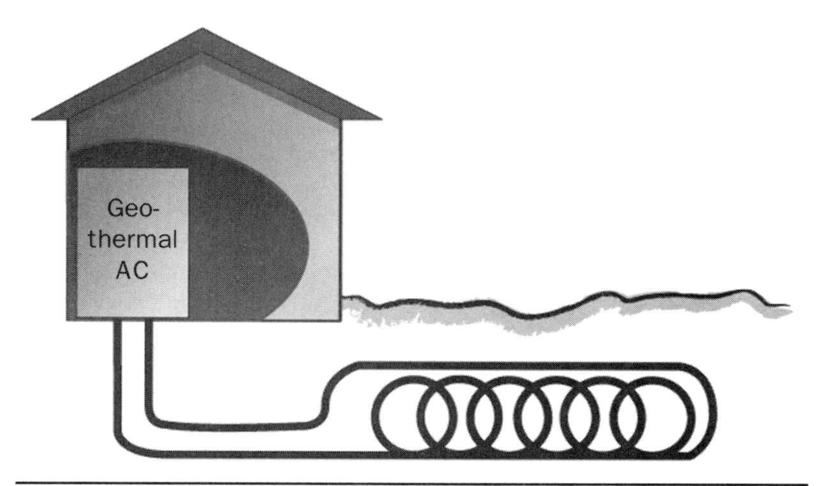

FIGURE 4-10 Of all loop designs, among the most commonly used in horizontal applications in the last 20 years is the "slinky" style ground loop exchanger. (*Sarah Cheney/Egg Systems*)

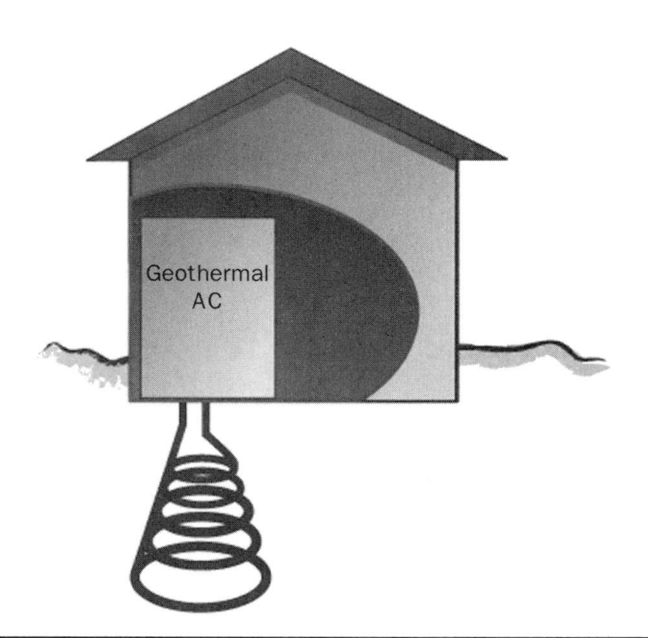

FIGURE 4-11 Slinky loops can be installed in a skinny trench standing up, as in Fig. 4-10, or in a 3-feet wide trench lying flat on the bottom, as pictured here. (*Sarah Cheney/Egg Systems*)

than dig several trenches, and install the successive loops of what is known as a *mat loop* (Fig. 4-12). As you can see in the picture, it does indeed look like a large black mat. Mat loops typically use 1000 to 1200 feet of piping per ton.

The slinky style of earth loop uses the available square footage to the best advantage of any of the horizontal systems. Loop contractors like to get the trenches as deep as is reasonably possible to be in damp, stable soil conditions, although there are drawbacks to this approach. For one thing, OSHA regulations require *shoring* or sloping for trenching at specific depths and intervals (though this seems to be rarely observed or enforced). Shoring is the reinforcing of trenches with metal or wood, to prevent the sides from caving in. This is a concern because working in trenches is among the more dangerous aspects of construction, and a cubic yard of soil can weigh up to 3000 lb (1360 kg)—enough

FIGURE 4-12 A mat style slinky loop is situated by laying the slinky assemblies flat on the bottom of the trench. In this photo, groundwater can be seen filling the trench prior to backfilling. (*Photo by Egg Systems*)

to kill a man in a collapse. Another issue is that although saturated soil is denser and conducts heat better, thermal retention becomes a problem in an area in which the seasons are too far out of balance. We'll talk more about this below when we discuss *degree days* of heating and cooling.

One potential problem to be aware of is that loops of a vertically positioned slinky can be easily crushed and kinked by a boulder, or by compaction before all of the voids are filled. Care must be taken during the entire installation process.

A *single-pipe horizontal loop* is generally considered the easiest and least invasive earth-coupled system to install. It requires the least labor and is generally what contractors recommend if there is enough space available. With a single pipe, the generally accepted practice is 500 to 700 feet of 3/4-inch HDPE per ton of residential heat absorption or rejection. Figure 4-13 illustrates an example.

The next easiest options are *two-pipe* (as in Fig. 4-14) and *four-pipe* horizontal configurations, which require a bit more labor. In all these cases, the loop contractor typically uses a chain trencher or a vibratory plow, and trenches through the property until the prescribed linear footage of piping is installed. Backfilling is relatively simple and often goes fairly quickly, using the blade on the front of the trencher to push the dirt back in the trench. A foot or more of earth is usually added before laying the next pipe. Rocky soil may require that extra backfill be brought in; otherwise, the piping can be kinked or severed, causing an expensive and time-consuming repair.

Keep in mind that horizontal loops generally require a larger section of property than vertical loops.

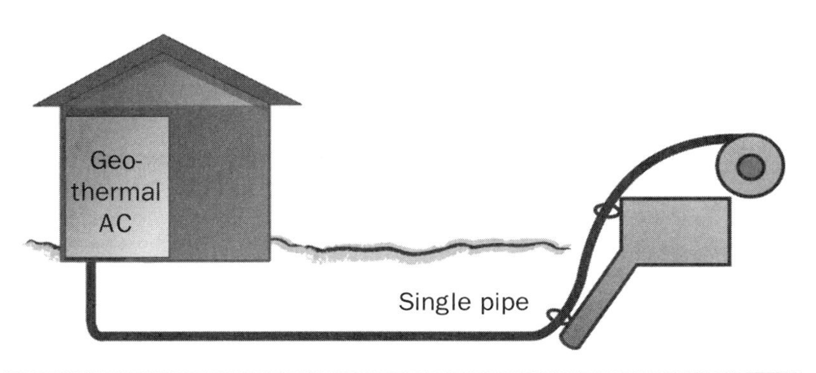

FIGURE 4-13 A vibratory plow attachment on a tractor allows for a single pipe to be installed in a trench that closes back on itself. (*Sarah Cheney/Egg Systems*)

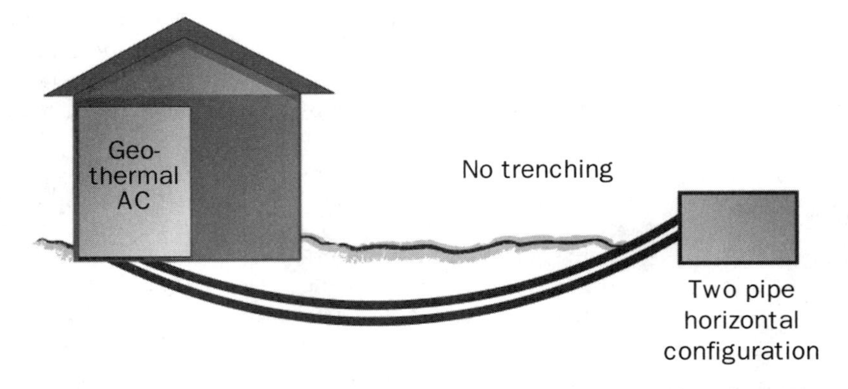

FIGURE 4-14 Horizontal boring has become increasingly popular over the last 20 years. Typically, two pipes can be pulled back through a small tunnel drilled by a horizontal drill rig. These rigs were developed primarily to allow for piping under obstacles, such as streets and buildings, without trenching. (*Sarah Cheney/Egg Systems*)

Pond, Lake, or Ocean Loops (Use with Caution)

Loops that are placed in a body of water, and not buried under the soil, are typically called *pond loops*. These can be placed in ponds, of course, or in lakes, oceans, sinkholes, streams, rivers, inlets, or anything else you may wish to use as a heat sink or source, as long as it is large enough to handle the load placed on it. These systems can be made of polyethylene pipe, or they can be metal exchangers called *lake plates* (see Fig. 4-15).

Contractors and homeowners will often assume, mistakenly, that a pond loop is a highly efficient system, when often the body of water the loop is placed in is quite subject to ambient outdoor temperature conditions. The efficiency is, therefore, affected by the weather in much the same way as an air-source air conditioner. When the air is 95 degrees, the water might be 89 degrees. When the air is 40 degrees, the water may be 55 degrees.

Another common problem is seasonal or periodic drops in water levels, or even freezing. I have heard too many times the response that the "loop will cover over with muck, and will transfer heat just fine." This is not true, and it will often come back to cause problems that will have to be addressed, during droughts that always come from time to time. Much like the earth, a body of water may need to be 12 to even 25 feet deep or more before it holds an average temperature well enough to be considered a highly efficient heat sink or source.

FIGURE 4-15 A Slim Jim lake plate being lowered into place by AWEB Supply. Plate exchangers have good heat transfer characteristics. (*Photo by AWEB Supply*)

In bodies of water with boat traffic, a loop is more likely to experience damage from time to time, unless they are completely protected by a dock or something of that sort. I have replaced or repaired dozens of loops that have been damaged by boat anchors or boat propellers at low tide. Many of these have even been completely underneath a dock. Most of the time, they are in inlet canals that have heavy boat traffic.

That being said, a major advantage of pond loops is their easy installation and low labor costs (Fig. 4-16).

Pumping Groundwater, Lake Water, or Seawater (Open-Loop Systems)

When I went to Oklahoma State University to be certified as a ground-loop designer in the early 90s, I was concerned by all the talk about closed-loop technology. I had had great success in my home with a system that used irrigation water to cool the air conditioner before doing its job of watering the lawn and bushes.

After I had been in training for two days and had not heard anything about such open systems, I asked Jim Bose what he thought

FIGURE 4-16 Pond loops are usually created by placing loops of HDPE pipe into the water, attaching weights, and then filling the pipe with water. Ponds and lakes follow ambient temperatures more closely than most would think...adequate depth is important. (*Photo by Egg Systems*)

about it. He answered that pump and reinject was a fine system as long as the geothermal air conditioner's exchanger could handle the water quality. I felt encouraged and resolved to leave my home system just like I had it at the time. It's been a great system and has only ever had one problem, which I will address in the section on scaling and fouling of exchangers.

As with closed-loop geothermal systems, open-loop technology can be unforgiving. An open-loop system typically pumps water from an aquifer, or an open body of water, and runs it through the heat pump to exchange temperature with the refrigerant inside the unit (see Fig. 4-17). The water is then discharged. Typically, it is recommended by the governing water authority to discharge the water back to the original source. It is not uncommon to see the water discharged through an irrigation system as reclaimed water, or dumped into an adjacent pond or canal after having been pumped from the aquifer. This is commonly called *pump and dump* and is generally considered a no-no.

Quite often, there is no monitoring of water consumption, and seemingly innocent pumping of water for air-conditioning can lead to millions of gallons of water waste, as well as the potential for depletion of the aquifer, or even a sinkhole type of event. A typical home

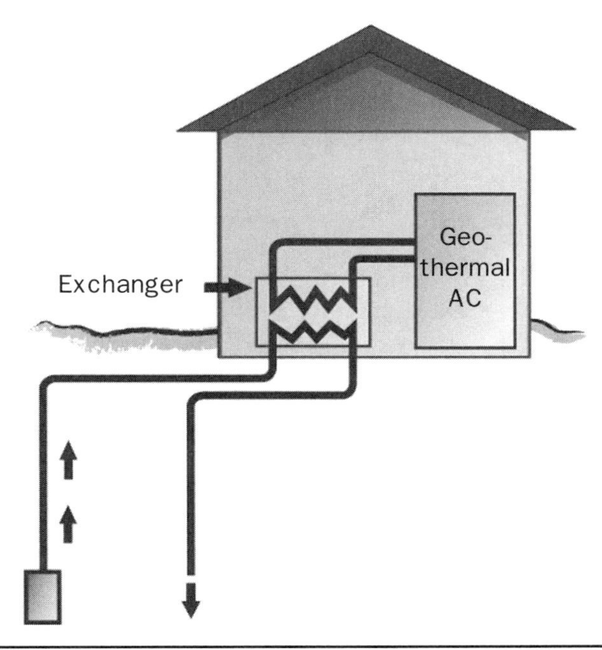

Figure 4-17 Open-loop systems may use an intermediate exchanger to protect the equipment from fouling due to water quality, and for other reasons, including enabling of load-sharing technology. (*Sarah Cheney/Egg Systems*)

geothermal system requires 6 to 8 gallons a minute when the system is running, which adds up quickly. Therefore, care must be taken.

Pump and Reinjection

In most cases, the best method for handling an open loop is with a so-called *pump and reinjection* system (Fig. 4-18). This is also called a *nonuse water well* system. In pump and reinjection, two water wells are drilled in the same vicinity. One is designated the supply well and the other is the injection well. This can be alternated for redundancy in lead and lag situations. This is a logical sequence that is implemented when two or more devices are cooling and heating a load. In order to foster equal wear and operation time, the lead-lag controller will alternate the #1 system designation. Otherwise, the #2 system might not operate for extended periods of time, placing an inordinate amount of wear and tear on the #1 system, and neglecting the operation of the backup or "peak" system.

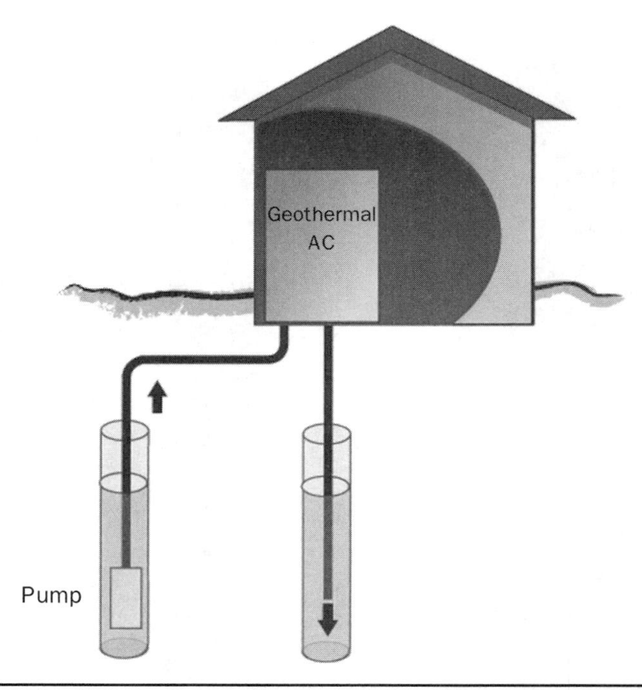

FIGURE 4-18 Open loops can also be called pump and reinjection when the water is returned to the same aquifer from which it was taken. This is the most environmentally friendly way to do an open system. (*Sarah Cheney/Egg Systems*)

Standing Column Wells

There is also a single well system called the *standing column well*. These must be very deep, sometimes several hundred yards deep. The water is drawn from the bottom and then returned to the top, where it percolates through a layer of added gravel, as illustrated in Fig. 4-19. One potential concern with this type of system is short-circuiting of the water from the return to the supply. Borehole testing is often needed to monitor this. There is ongoing testing of this technology, but right now, drilling costs are often prohibitive.

Surficial Aquifers and Caisson Infiltration

Where the soil percolation is good, as in sandy areas, *surficial aquifers* offer an easy answer to the problem of water availability. Most often, these areas are tropical or subtropical in nature and have abundant rainfall. When this is the case, a shallow well can be installed, usually under 30 feet deep. The sandy conditions allow the groundwater to

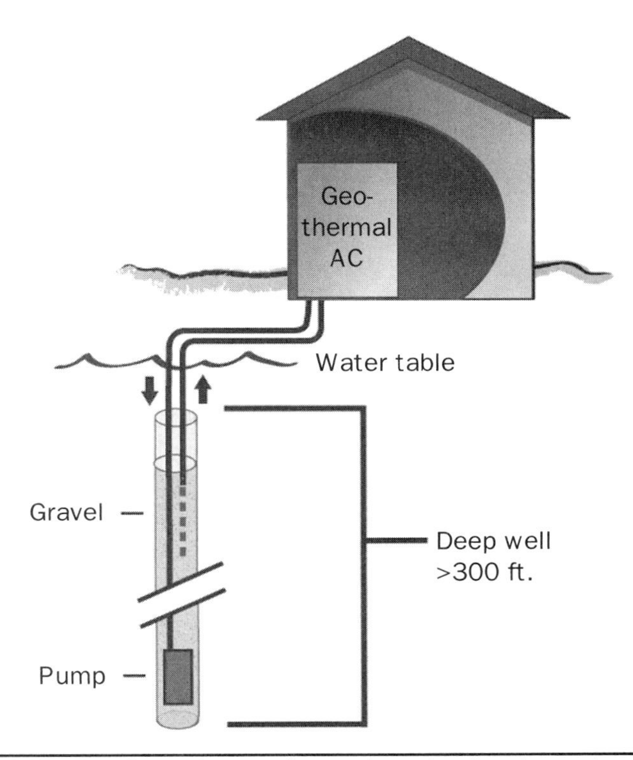

FIGURE 4-19 Standing column wells are large in diameter and sometimes several hundred yards deep. Return water is percolated from the top through layers of gravel and pumped back up to the equipment it serves. (*Sarah Cheney/ Egg Systems*)

percolate through the granules quickly and easily as water is pumped out. While this is usually a good source of fair-quality water for earth-coupled purposes, it is not as dependable as deeper sources. A drought year can quickly have a negative effect on the ability of a shallow well system.

Caissons are large-diameter holes (24 inches or more) bored into the earth. They can be 10 to 30 feet deep and are filled with a large aggregate, allowing good flow in and out for the purpose of pumping and returning water (see Fig. 4-20).

Concerns with Open Loops

You may ask why we stated that open-loop technology is so unforgiving when it may seem that we only have to pump water and inject it back in. We should clarify by pointing out that problems could begin to manifest themselves after only a few short years if installed improperly. Over the years that I have been involved with this

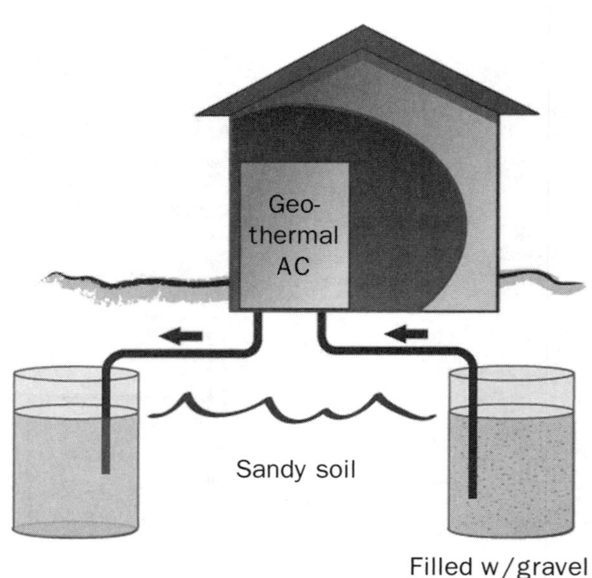

FIGURE 4-20 Caissons drilled into sandy soil are used for supply and return in areas with high rainfall and shallow aquifers. (*Sarah Cheney/Egg Systems*)

technology, I have seen many water exchangers fouled with scale and debris. It's important to test the water before you start and make any adjustments as needed.

There are methods of piping and engineering these pump and reinjection systems that can increase their performance. The introduction of new air into the system should be eliminated as this promotes fouling. This can inadvertently occur through leaky suction fittings on an above-ground pump system. The velocity rated for the exchanger should also be maintained. Furthermore, ensuring that the injection piping goes below the surface of the water in the injection well helps promote a siphoning effect (Fig. 4-21), reducing the power needed for pumping and increasing performance.

In systems with multiple loads, the water should be pumped through an intermediate exchanger that is easily serviceable, making cleaning easier. Some manufacturers are introducing self-cleaning exchangers with enough turbulent flow to affect favorable heat transfer rates while promoting lower maintenance. The use of an intermediate exchanger allows the loads to be on a closed loop, which eliminates the need for cleaning of the individual exchangers on the secondary loop, and allows more complete load-sharing operations.

FIGURE 4-21 By using the same principle as siphoning out a fish tank, deep aquifer supply and injection wells can use the siphon effect to reduce pumping power. The drawdown reduces the energy needed for the pumping process. (*Sarah Cheney/Egg Systems*)

Ted Chittem of Connecticut's Air Perfect, Inc., told us that he thinks open loops require a

> higher level of competence versus closed loops. [He continued,] and you have a concern about mineral content that you wouldn't have if you were doing a closed loop. So it's necessary to test the water before you commit. You may need to upgrade from standard copper heat exchange material to nickel alloy, which is less susceptible to damage from minerals. You have more maintenance with open loops. It's a tradeoff.

Open loops are prohibited in some parts of the world, largely out of fear of depletion of aquifers. While this won't happen with properly designed pump and reinjection or standing column systems, it is important to be aware of the issue. Groundwater can be depleted faster than you might think; a typical 5-ton geothermal air conditioner can pump a quarter of a million gallons of water per year. Chittem pointed to a hotel he was familiar with that ran what he called "pump and dump" with a geothermal air conditioner for 25 years, diverting used water into a drain. "They dropped the aquifer a foot, and that's water that took hundreds of thousands of years to fill," he told us.

Aquifer depletion can cause a number of problems. It can cause wells to run dry and can threaten natural springs, ponds, and rivers. It can also lead to the formation of sinkholes or other property damage and possibly even loss of life. Groundwater depletion can also degrade natural ecosystems.

In many parts of the world it is illegal to drill a water well for consumptive use if you have access to city water systems available. But in many of those places, you can drill a nonconsumptive use well, such as for earth-coupled HVAC. Often the codes mandate that two identical holes be drilled. This also lends itself to redundancy, which is important in commercial applications where a backup system is needed. Each well has a pump and a discharge pipe allowing the backup systems to be employed at any time.

In some places, you can't drill into certain aquifers for any reason, so it's important to work with installers who know the local area. Such restrictions may change, however, as awareness increases and the geothermal market gains acceptance. As an example, Florida does not allow drilling in many areas without a casing to seal the hole. This adds significant expense to the process of drilling deeper than 30 feet. The option of drilling several little holes, however, is expensive due to the increased labor and the addition of more fittings for the pipe.

There is an argument that pumping from a deep well is an energy deficit. It is true that a deep well pump is going to use more power than a closed-loop recirculation pump. But the Air-Conditioning, Heating, and Refrigeration Institute (AHRI) testing criteria that rate groundwater air conditioners take this higher pumping power into consideration. Taking this a step further toward energy conservation and efficiency, proper piping of a pump and reinjection system includes placing the dip or return pipe fully into the water table of the return well. This allows the siphon effect to take place, much like how siphoning out a fish tank works. This significantly reduces the pumping power needed for the job.

With reduced pumping power, and the implementation of a pump with a variable frequency drive (VFD), the overall benefit is increased. A one-horsepower pump draws 1000 W or more (up to 3000 W, depending on the type of motor), but a VFD motor (pictured in Fig. 4-22*a* and *b*) draws only what is needed at the time, which I can tell you from experience is often 25% of the standard motor load. Imagine dropping from 1600 W to just 400 W for pumping power. ... It really works!

FIGURE 4-22 (*a*) A variable frequency drive (VFD) gives a 3-phase AC motor the ability to vary its speed by adjusting the pulses or frequency of the alternating current to the motor. This technology is another key to reducing pumping power while allowing for multiple load factors. (*b*) Another VFD.

Open Loops versus Closed Loops

In April 2010, a reader posed a question on RenewableEnergyWorld.com that has worried me for at least the past 17 years. Someone asked Scott Sklar, president of the Stella Group—a marketing firm focused on distributed energy technologies—whether closed-loop systems "only last about 10 years because the ground temperature equalizes after several years as a result of so much heat being removed." Sklar answered that a brief survey of the major geothermal HVAC manufacturers turned up no reports of this kind of saturation problem for residential systems. He talked to Dan Ellis, president of Oklahoma-City-based ClimateMaster, the world's biggest producer of geothermal HVAC systems. Ellis told Sklar that problematic heat retention could be seen "only in larger commercial systems with a dense ground heat exchanger array and with unbalanced seasonal loads (i.e., more heat rejection than extraction or vice versa)."

Ellis said geothermal design software accounts for this potential issue, and helps the installer set up a system that will work over long periods of time. "Alternatively, the load can be balanced with supplemental heat rejection or in many others ways," Ellis told Sklar, who added, "I want to emphasize that even for larger commercial systems, geothermal heat pump systems and hybrids with solar thermal are the most energy efficient systems for heating and cooling buildings. Overall, I am a big fan."

Ellis is right about the fact that commercial systems often have imbalanced seasonal loads causing thermal storage or thermal retention. But what is often missed is that this also happens in residential closed-loop applications throughout the south.

In 1992, Egg Geothermal installed a system in a home in Tarpon Springs, along the west coast of Florida, that is a textbook example of a geothermal slinky-style mat loop. The loop was installed between seven and eight feet deep in pure sugar sand, below the mean surficial water level. This type of soil is considered among the best of thermal conductors. I oversized the system by 12% according to known industry standards and it performed well the first year. The second year, we began to experience thermal retention with higher-than-normal return temperatures. By the third year, the retained heat actually killed almost an acre of this poor man's front lawn.

This is not a singular incident. This has happened dozens of times to me, before I put a stop to it by refusing to engineer systems of this nature. However, with the resurgence of geothermal air-conditioning installations, it is happening again with a new group of contractors. In April 2010, my sales team traveled to Orlando and attended a sales meeting for Indiana-based geothermal HVAC manufacturer WaterFurnace, and they reported to me that closed-loop technology was the only type of installation being taught. After just a few years of operation in the hot and humid southern regions, where ground temperatures are in the 70s, many closed-loop systems will not perform any better than standard high efficiency air-source heat pumps. This issue is starting to be addressed, but consumers and contractors must be educated.

In the proper climates and soil conditions, closed-loop technologies of all kinds have their place. My brother Brian Egg is the president of Egg Systems Southeast, based out of Atlanta. He reports that their closed-loop activity is brisk with no signs of thermal retention in either vertically bored systems or horizontal loops.

Part of the secret is properly sizing the ground loop for the heating loads of the building. It also can be helpful to use a fluid cooler to supplement the heat rejection in the cooling load. This can often bring down the initial cost of the system, too, although it does add to the complexity of controls, and adds a piece of outdoor equipment, along with water treatment considerations. However, the potential for cost reduction in cooling-dominant loads may sometimes justify this option.

In their training course to engineers, major HVAC manufacturer Trane states that, "The use of groundwater directly in the heat pumps, however, results in the best overall system efficiency due to the constant temperature of the water. This can also have the lowest installed cost of any ground-source system." The training goes on to mention that in certain situations, it is recommended to "help ensure the water flowing through the heat pumps is clean, an intermediate plate and frame heat exchanger is recommended." This is the reason that, in order to have a properly designed geothermal system for your installation, a highly qualified geothermal systems engineering firm should be retained. Well-engineered systems can operate with as little as a 3°F approach, meaning the difference between well temperature and loop temperature is only 3°F.

Open-loop designs are almost always more efficient than any other type of system because they are better coupled with the ground temperatures. Closed-loop systems, in comparison, have to transfer heat through more layers of polyethylene piping and the soil.

There are many who claim that using open-loop technologies are out of the question because of the losses in efficiency due to pumping power. This is simply not true.

Smart-Energy, a geothermal installer in Queensbury, New York, argues,

> In the Northeast, over the past 30 some years we have found geothermal units operating under AHRI-325(50) [open loop] to be significantly more efficient, and have higher output capacities, than their counterparts operating on AHRI-330 [closed loops]. The argument that closed loops are more efficient in AHRI-330 because they use circulators wherein AHRI-325(50) we use pumps has been proven wrong consistently by ARI [Air Conditioning Research Institute] standard tests.[1]

Why? According to Smart-Energy, an open-loop geothermal system produces higher levels of both heating and cooling than the same unit operating on a closed loop, thereby requiring less run time to condition the space. "Given the advantage of the clean water (fuel) source and the high density bedrock we have throughout the Northeast, why would anyone throw away this clear performance advantage and opt for a closed-loop versus an open system?" asks Smart-Energy.

To gain a bit more understanding, here's a look at AHRI's ratings:

- *AHRI-320*: Not a geothermal standard, but refers to closed-loop, water-source HVAC units that work with a boiler and cooling tower. In this case, performance numbers do not include the required pumping energy costs to supply the units with water or fluid, or the fan power of the cooling tower.

[1] Note that in 2007, ARI merged with the Gas Appliance Manufacturers Association, to form the Air-Conditioning, Heating, and Refrigeration Institute (AHRI).

FIGURE 4-23 Jay Egg (left) and Don Goldstein inspect the pump for Mr. Goldstein's new geothermal system. (*Photo by Egg Systems*)

- *AHRI-325(70):* This is for open-loop geothermal systems where average groundwater temperatures vary around 70°F (predominantly in the south). These ratings do include pumping energy costs to supply the units with water.
- *AHRI-325(50)*: This is for open-loop geothermal systems where average groundwater temperatures vary around 50°F (predominantly in the north). These performance figures also include pumping energy costs.
- *AHRI-330*: This is for geothermal equipment operated on a closed loop. The figures also include pumping energy costs, this time to supply an exchange fluid, typically antifreeze. The standard input for heating is 32°F, and for cooling is 77°F entering fluid temperature.

Egg Geothermal finds that while closed-loop systems have many good applications, it is rare that an open-loop pump and reinjection install would not be a more efficient opportunity, especially in commercial and cooling-dominant situations. The variables, however, are extraordinary, and there is no set rule of thumb that can be followed. Each system must be custom engineered for proper heat transfer and fluid flow so that maximum efficiency is attained. To guess at this type of thing can cost thousands and even millions of dollars over the life of a system. Something as simple as a $10,000 modification in exchanger choice or pumping arrangements can save hundreds of thousands over the life of a system, extend life, and reduce maintenance costs.

Summary

We have discussed the major types of ground loops, from horizontal to vertical and from open to closed. Hopefully, the reader has gained a sense of the advantages and disadvantages of each design and is armed with questions for contractors on how best to meet the demands of the site. No job is the same, but the good news is that an increasing range of solutions are available for qualified contractors to get good results from each project. Earth-coupled systems tend to deliver steady, even heating or cooling at optimal temperature and humidity. They can work well with vent systems or with radiant heaters that circulate hot water under floors, producing an efficient, allergen-free warmth.

CHAPTER 5

Introduction to Load Sharing

Right after I got out of the Navy in 1987, I was hired by Orlando Foods, a company that operated 25 Wendy's restaurants in the Orlando, Florida, area. During the year and a half I was employed there, we opened two or three brand new stores, and did several remodels. At a particular store on the north side of the city, I got to install perhaps the first heat pump water heater anyone had ever seen up to that time in the area.

There were many who thought this was about the dumbest thing they had ever seen or heard of... an entire refrigeration system just to heat water, when gas or electric was all that was needed? The payback on investment was about four years as I recall. The heat needed for the restaurant's water was absorbed from the food preparation area in the back room, where the dishes were washed, the chili was cooked, and the potatoes were baked. An evaporator, or cold coil, hung from the ceiling with a fan to draw the air in the back room through it, as in Fig. 5-1. The result was cool conditioned air to take the edge off of the heat produced by the dishwasher, ovens, and chili pots in the back room. All of the hot water needed for the restaurant was produced by this system under ideal conditions.

This is a perfect example of load sharing. Load sharing in commercial and residential applications becomes as limitless as heating and cooling loads themselves (Fig. 5-2). Below, we'll review some of the loads that can be considered.

Benefits of Load Sharing

As we saw in Chapter 1, air-conditioning is often the biggest energy user in a typical building. The next biggest energy users are often water heating, refrigeration, or pool heating and pumps. Lighting and small appliances are lumped into the third place spot. In hotel and restaurant applications, hot water generation and pool heating are significant opportunities for reductions in energy costs. Use of geothermal technology

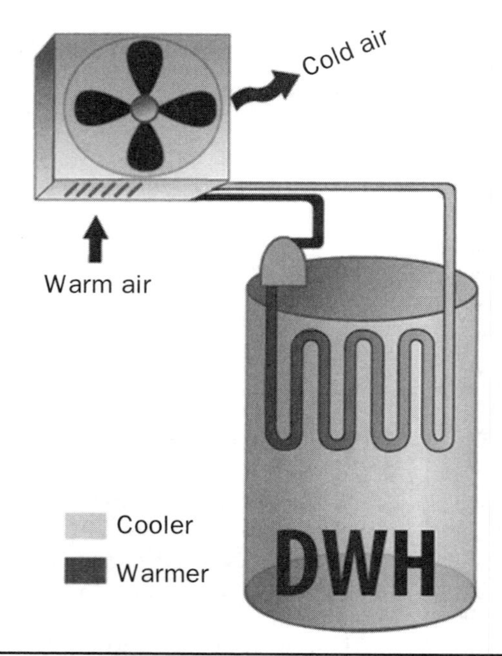

Figure 5-1 A heat pump water heater provides cool conditioned air as a free byproduct of water heating. Geothermal systems can capitalize on load sharing in much the same way. (*Sarah Cheney/Egg Systems*)

in these cases can offer 70% to 90% savings when compared to the costs of gas or electric equipment.

Beyond these, there are countless other opportunities in commercial applications. Cooling towers, which currently dot the rooftops of skyscrapers and occupy valuable ground space of school complexes, can be eliminated. The consequences of this single upgrade are many. Billions of gallons of freshwater are evaporated into the air, and millions of dollars are spent on chemicals, to control water quality and prevent scaling and calcification of the surface of the cooling tower and associated heat exchangers. The primary purpose of a cooling tower is to take condenser water and cool it down to a usable temperature for the refrigeration and air-conditioning equipment it services. They are loud and require endless maintenance, repair, and replacement. By pumping and reinjecting groundwater, instead of using the evaporative cooling effect of a cooling tower, water waste will be all but eliminated and the efficiency of the refrigeration equipment will increase, saving substantial amounts of energy.

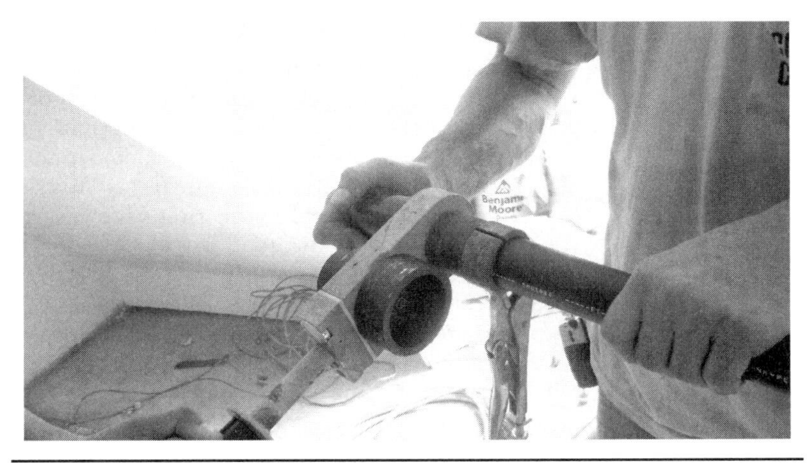

FIGURE 5-2 An HDPE coupling is joined to a section of the pipe using the heat fusion process, guaranteeing no leaks and a long life. This piping is used in load-sharing circuits and ground coupling of geothermal HVAC systems. (*Photo by Egg Systems*)

The Case of a Large Hotel

By now, it should be clear that air-conditioning systems are a perfect candidate for earth-coupled applications. Let's go over some other uses in a little more detail, using a large hotel as an example. First, there are likely lots of air conditioners in a typical large hotel, even hundreds of them. Then, each room uses considerable amounts of hot water for each tenant, both for showers and for bathing, as well as towels and linens. A heated pool may be included. The energy, either electrical or gas heat, to keep a pool warm amounts to tens of thousands, and sometimes more than a hundred thousand dollars per year (really!). Refrigeration equipment, such as ice machines, walk-in coolers and freezers, and ice cream-making equipment, are all putting out heat.

Take a moment to think about how many systems need heat and how many give up heat. Have you noticed how *hot* the little vending machine niche is from which you retrieve ice in a hotel? That heat is not only wasted, but causes discomfort. Then, after retrieving your ice, you go take a hot bath or shower that steams up the room, causing the air conditioner to have to remove the heat and humidity. All the while the pool heaters are running downstairs, using hundreds of thousands of units of heat by burning fuel. Let's follow a unit of heat energy through two different scenarios in a hotel.

Scenario 1

As in Fig. 5-3, a customer comes into the room and turns the air down. . . . Cool air comes into the room; hot air is wasted to the outdoors. (1) He gets some ice and complains about how hot stupid little vending rooms are (hot air wasted), and hallway AC load is increased to remove the waste heat. (2) He comes back to the room and takes a shower for 10 min and uses a couple of towels; hot water heat and steam increase load for the AC in the room, which works harder, not to mention a higher hot water load is created by the towels for the hotel laundry. (3) Domestic water heaters (and pool heaters) are burning gas to heat the water; and the waste ventilation heat is going up the vent pipe and to the outdoors (not to mention the carbon footprint of CO_2 emissions).

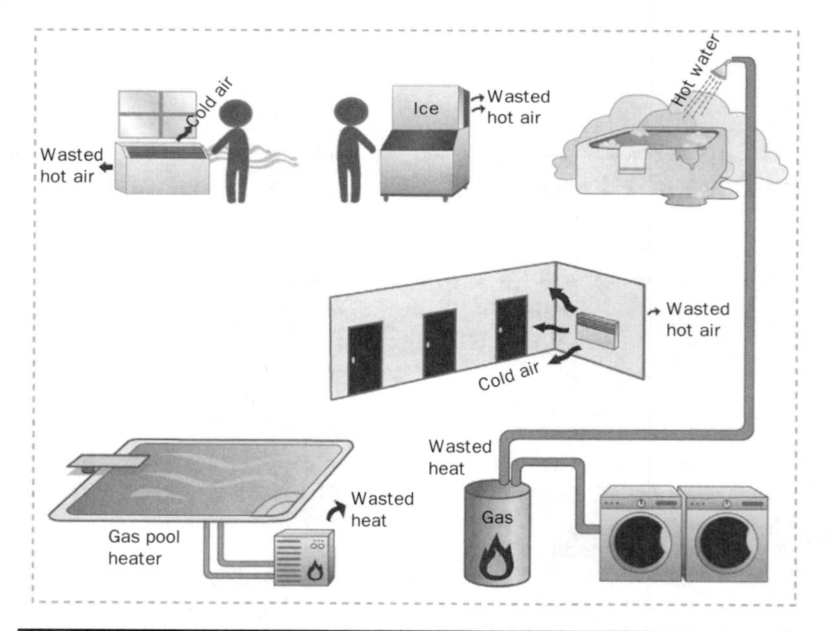

FIGURE 5-3 Many homes and commercial buildings have several areas of wasted energy that could be otherwise harnessed and reused. In this example from a hotel, some of the wasted energy includes (1) waste heat from the room air conditioner, (2) waste heat from the ice machines, (3) waste heat from the shower steam, (4) the need for energy to heat water for laundry and showers, (5) the need for energy to heat water for the pool, (6) waste heat from common area cooling for hallways and foyers. (*Sarah Cheney/Egg Systems*)

By now, we see the incredible waste of energy in any hotel, or almost any building for that matter. But the question is: How can we get the waste heat from all of these processes to one another? Should we run a complex duct system to each room air conditioner to "collect" the heat being wasted from the room AC? Perhaps the same with the ice vending machines? And what do we do with it? Blow it through the water used for laundry and showers, and perhaps even the water used for the pool? Not only will this *not* work the way that I have just lamented, but also it would be miserably expensive and woefully ugly.

Scenario 2

In the scenario in Fig. 5-4, all of the equipment discussed above is now water-cooled refrigeration equipment, often referred to as *water-cooled heat pump equipment*, meaning it can reverse its cycle for the load, either

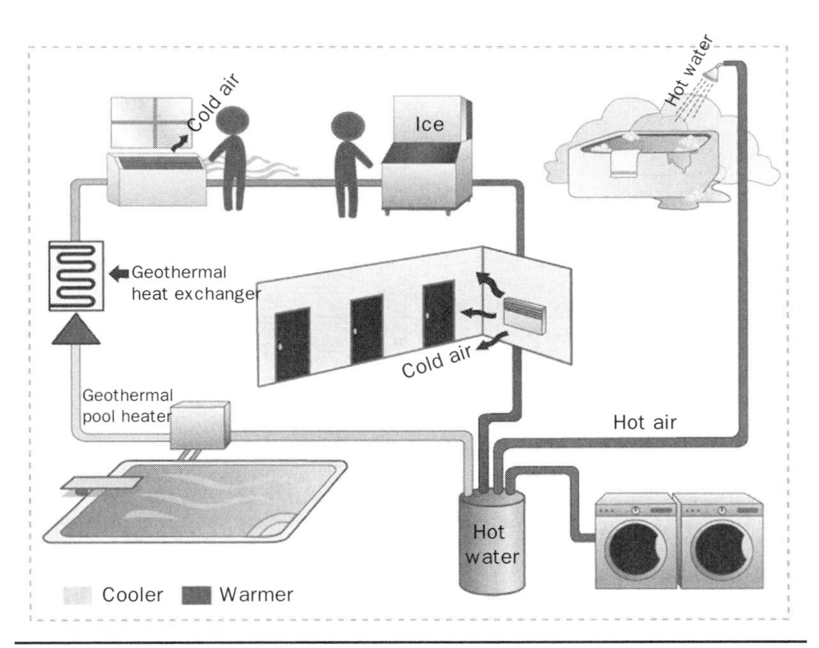

FIGURE 5-4 Utilizing load sharing through a primary geothermal loop, the individual components of a geothermal facility are in harmony: The waste heat from various processes is used to help heat water for swimming pools, laundry, and other uses. The result is near-zero energy consumption for many of the components and drastic reduction in carbon footprints. (*Sarah Cheney/Egg Systems*)

pumping heat into the load or extracting heat out of the load. Remember from our lesson on thermodynamics that liquids are supremely more efficient in transferring heat than air.

Now we are ready to follow our hotel patron from scenario 1 above. The customer comes into the room and turns the air down. Cool air comes into the room—the result from his water-cooled AC is that warm water is extracted. (1) He gets some ice and notices there are no noisy fans blasting hot air into the vending niche. A smile crosses his face, and the hallway AC load is unaffected by the ice vending equipment. (2) He comes back to the room and takes a shower for 10 min and uses a couple of towels; hot water heat and steam increase load for the AC in the room, which works harder to remove the heat and steam, not to mention more hot water load created by the towels for the hotel laundry, resulting in more warm water extracted from his room AC. (3) Domestic water heaters (and pool heaters) are no longer burning gas, since they are now using the warm water discharge from the room AC, the ice vending equipment and the hallway AC systems, putting that heat energy toward the jobs of making *domestic hot water* and warming the pool water (through

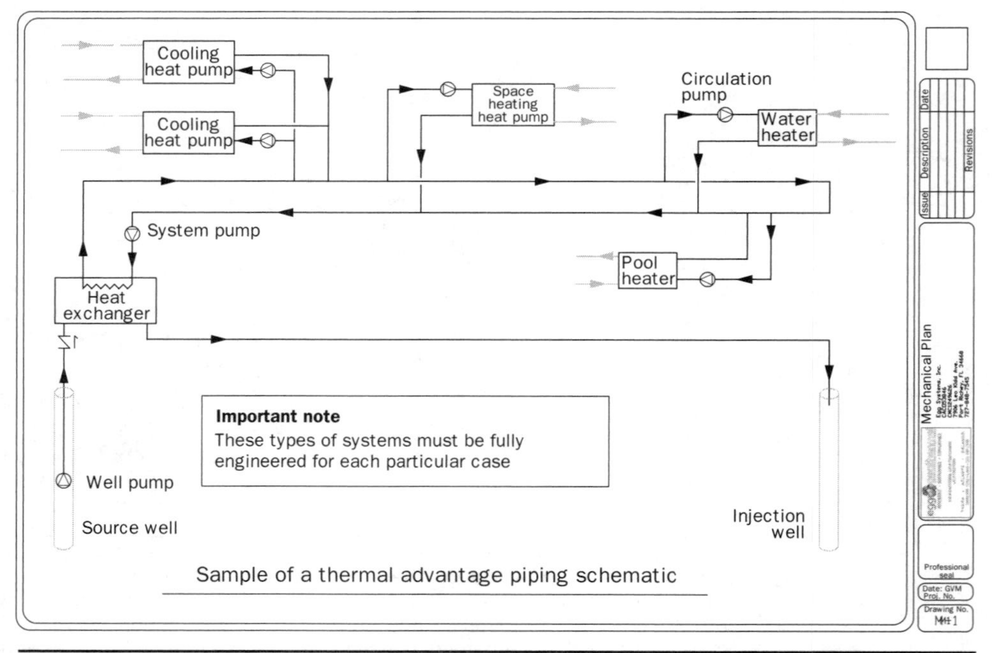

FIGURE 5-5 Engineered drawings for load sharing may seem complex, because they are. If not properly designed, the result can lead to system failure or even catastrophe. Always use an experienced professional engineer for your geothermal systems. (*Guy Van Meulebrouck, PE/Egg Systems*)

their highly efficient, water-cooled dedicated heating equipment). There is no burning gas to heat the water, no waste ventilation heat going up the vent pipe and to the outdoors, and no carbon footprint from burning additional fuels.

Just for fun, you can turn the above scenario around in your mind and consider the resulting cooler water extracted or discharged from the pool and domestic hot water systems. This is piped to the room air conditioners, providing a much-appreciated boost in their performance.

This is the essence of load sharing (see Fig. 5-5 for an example of an engineering drawing for load sharing). The question has always been how to exchange one with the other? The answer is clear: liquid source or geothermal equipment.

Earth Coupling as Thermal Savings Bank

In the use of earth-coupled air-conditioning, the common exchange medium is the liquid loop. Sometimes the loop is absorbing heat, as in the process of cooling or freezing something. Or, the loop could be giving up heat, as in the process of heating water or heating rooms, or even parking lots or driveways. Properly piping the loop will enable a building to share all the loads of the building system, dramatically reducing the electrical consumption.

I grew up the oldest of nine children in the Mojave Desert of southern California, in a large home on nine acres. My parents were green before green was cool. And family was central to all we did. We were active in the community and fully engaged in self-sustenance and Provident Living, principles deeply instilled in us as members of the Church of Jesus Christ of Latter Day Saints (The Mormon Church). We found ourselves in the local news often for achievements or good works of one kind or another.

I fondly remember providing food and goods to local charities as a part of the Young Men's and Young Women's programs. A local contractor who owned an excavation company pulled us around in the bed of his "Monster Truck," named *Big Blue*. We all would run to a door and sing a couple of Christmas carols, then hand over bags of food and other items—such as blankets, scarves, and wooden toys—that we had been working on for the previous month.

When we moved into our desert home in December of 1974 from a neighborhood called Irwin Estates in Barstow, my father ran the heat for only one month. The energy bill was $160. I'll never forget how shocked he looked. To me, this type of energy bill was the epitome of "not sustainable." We simply could not "sustain" the payment of this high electrical bill. The only day heat was ever run in the bedroom wing again was on Christmas Eve—just one night a year. But make no mistake: The winters were long and cold.

Each of us had our own personal form of thermal savings: a large smooth stone, between 5 and 10 pounds, carefully selected from the rocky hills and dry streambeds behind our home. We would place them carefully atop the Franklin wood-burning stove that heated our front rooms. When bedtime came, each of us wrapped our stone in our own towel, which kept it from burning our hands, and hurried to bed. Even as a 16-year-old, I can remember climbing into that ice-cold bed and taking my stone out of the towel and sliding it up and down and around inside the covers. Slowly, I would slide in as the stone heated up the bedding. Inevitably, all of the heat would be gone within 15 or 20 min, but then good old body heat took over.

As I lay there, I remember wishing that I could find a way to store some of the brutal summer heat for those cold winter nights. Though the technology was not available then, we have it now.

Still, we have articulated the premise of earth-coupled thermal retention as a means of load sharing. I have actually never heard it taught or read of it anywhere. So here you go:

All winter long, a geothermal or earth-coupled AC unit is pumping heat from the ground to warm the building it is serving. Under most conditions, the result by the end of a long winter is inevitably going to be a "plume" of colder earth, as shown in Fig. 5-6. Much like a glass of ice water set on a table in the middle of a room, this plume (and the ice water) will normalize after a period of time. In the case of the ice water it may take several hours to assume ambient temperature. In the case of the earth temperature plume, it may be months before it assumes ambient soil temperature. If this seems worrisome in the least, consider this: The first time the cooling is turned on, the geothermal AC system has a colder-than-normal source of earth coupling because of the colder plume created by the long winter. The result is extraordinarily efficient operating conditions for the equipment because the source water is cooled to below the normal ambient earth temperature.

As the summer matures, the remaining deposits from the winter "cold" are used up, and the AC system begins to swing the other way, as it begins to heat the earth up above the ambient soil temperatures, creating a plume of heat, as in Fig. 5-7. Then, when winter comes, there in the earth are those wonderful deposits of heat from a long summer, just waiting to be extracted for use during the upcoming winter.

This great benefit of the load sharing is also the reason why some areas of the world are not conducive to closed-loop, earth-coupled technology. Whenever there is a stark imbalance of heating hours as compared to cooling hours, the project will suffer from compounded thermal retention. Unless the thermal conductivity characteristics of the earth in that region surpass the heat transfer rate, the system will

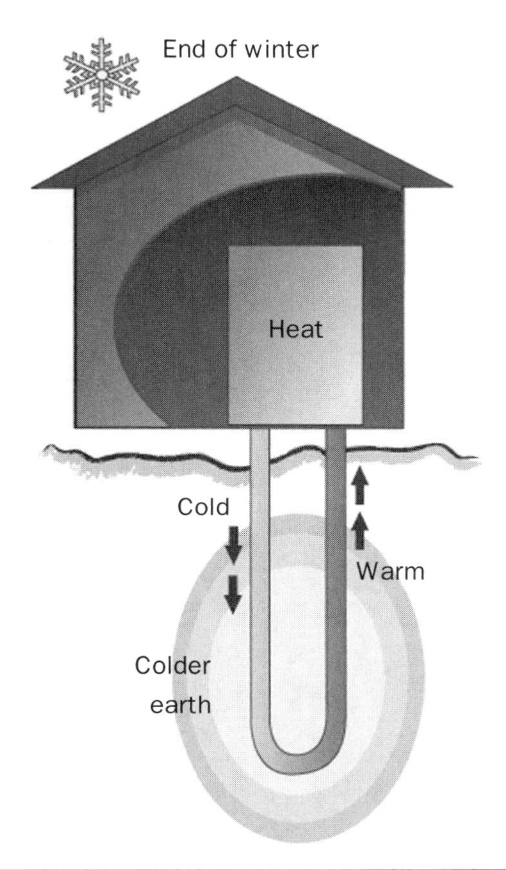

Figure 5-6 All winter long, heat is extracted from the ground, leaving a relatively cool plume or bulb around the earth-coupled ground loop. (*Sarah Cheney/Egg Systems*)

fail. To summarize, if you live in a very warm, or a very cold climate, you may be in a situation in which the geothermal technology that works best for you departs from normal earth-coupled practices and training. But rest assured, we interviewed installers who have had great success in chilly Alaska and the high desert of the American southwest. So stay tuned!

I live in Florida, and geothermal earth-coupled technology is thriving here. We pump the groundwater, and inject it back into the earth—this practice uses what is commonly called *open loop* or *pump and reinjection*

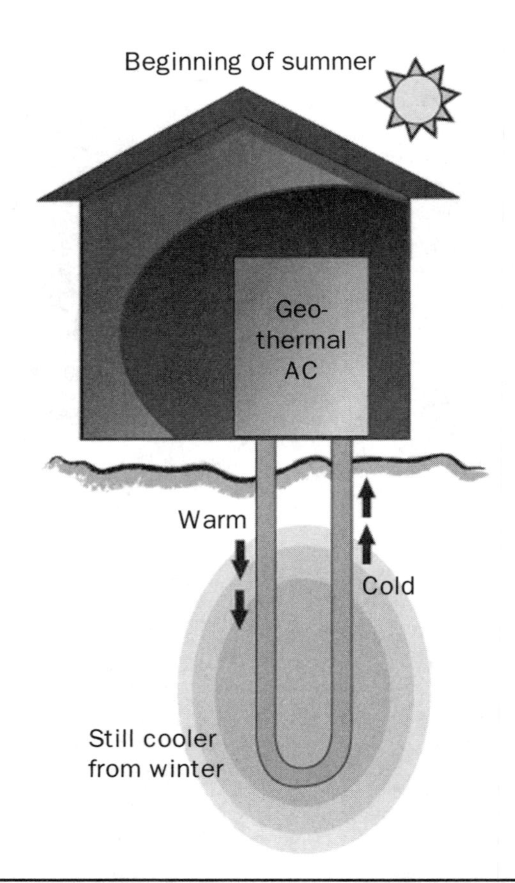

FIGURE 5-7 At the beginning of summer, the geothermal heat pump capitalizes on the relatively cooler earth temperature from the previous winter to help cool the home. (*Sarah Cheney/Egg Systems*)

technology and is known by most water-regulating authorities as a "nonuse well permit." *Nonuse* refers to the fact that all of the water is returned to the aquifer from which it came. (See Chaps 6 and 7 for more on this.) This practice, in and of itself, has a few intricacies that make it extraordinarily energy efficient, renewable, and sustainable. However, improperly installed, it can be a disaster.

IMPORTANT NOTE: *Always make certain your contractor has the experience needed for your region. If you're unsure, have the job engineered by a trusted geothermal firm.*

Summary

Has it ever bothered you that one device, say a refrigerator or an ice maker, produces excess heat that is simply wasted, while you have to pay good money to produce heat in another area of a structure, say a hot water heater or furnace? Well, the concept of load sharing can help address this imbalance by taking advantage of the heat differential of one device and channeling that into useful energy for another. Luckily, the ground provides a handy sink to facilitate this process, and earth-coupled systems can efficiently store heat energy for when you need it most.

CHAPTER 6

Efficiency and Load Calculations Simplified

During the energy crisis of the 1970s, I did what many did in the face of long gas lines and high prices: I complained a lot. I also bought a carburetor kit from a catalog that promised a remarkable savings on gas of nearly 60%.

We had been getting about 9 miles to the gallon in our Dodge 1964 wagon, so I was hoping for 15 miles a gallon. This carburetor had new jets and different throttle parts, as well as a nice looking new air filter.

After installation of the kit, I had to wait for my parents to fill the car up and reset the odometer, since I wasn't old enough to drive. After a few weeks, my dad handed me a note with the missing information filled in: 242 miles, 23.4 gallons. All that we had attained was a disappointing 10.34 miles per gallon.

I called the main offices where I had purchased the kit. After an hour on hold, I was told by one of their service specialists that we had not followed the guidelines. They stated something like this: "A 60% fuel efficiency increase may be seen with a combination of the installation of this kit and modification of driving habits. This includes driving on highways at speed not to exceed 45 miles per hour. Begin decelerations about 1000 feet before an intersection. Keep vehicle weight lower by removing unnecessary items. Keep tires inflated..." The list went on.

I was not happy with all of the conditions I had to follow for a product I had purchased in good faith. In fact, I felt that I had been lied to. I didn't feel that a person should have to change his lifestyle to achieve energy savings promised by a product that did not boldly proclaim that such a change was necessary. If that were the case, I would have purchased a guide to energy-efficient driving habits.

I'll bet you get the point of this story. Just in case, here it is: There are a lot of variables in energy efficiency, including with geothermal air-conditioning equipment. A contractor or a brochure can say that a system has an energy efficiency rating (EER) of 30. But what does that really mean? Certainly, you may be able to say that compares favorably to an EER of 15 or 19. And you can make a correlation to the types of air conditioners that are measured by seasonal energy efficiency ratings (SEERs).

IMPORTANT NOTE: *How do you know you are getting the efficiency a contractor says you are getting? This chapter will arm you with the ability to make that determination.*

Rating Geothermal Systems

As we stated above, the rating systems for geothermal equipment (Fig. 6-1) are a bit different than other residential HVAC equipment.

FIGURE 6-1 As Jay Egg examines a geothermal air conditioner installation with a client, he explains the benefits: (1) superior comfort, (2) unbeatable energy savings, (3) quiet operation, (4) three times the longevity due to the elimination of the outdoor portion of the system, (5) government incentives, (6) free hot water, (7) fits right in a closet or any other location. (*Photo by Egg Systems*)

SEERs take into consideration that the cooling portion of your bill takes place during different times of the year. For example, at high noon the temperature may be 79 degrees in April, but 95 degrees in August. The instantaneous rating for a system may range from 20 EER to 10 EER, settling in at a seasonal rating of 15 SEER. This is oversimplified, but you hopefully get the idea. The reason that geothermal or water-sourced equipment is rated without the seasonal factor is related to the stability of the source or fluid temperature. Typically, a properly engineered system will maintain a fairly constant source temperature. If a system is designed on a loop that has low capacity, you may experience large fluctuations in liquid temperature that can affect energy efficiency, and in some cases may even stop system operation. Let's look at some simple formulas to determine your EERs.

Before you get into this section, we want to clarify that there is a good reason for putting you through this exercise. If it seems too daunting, hire an independent consultant to do it for you. I have seen many great systems installed incorrectly, causing customer complaints and comments such as, "My bill is still the same as last month or last year at the same time," or, "I paid all this money and I really don't see the payback coming in." Often times, the contractor will shoot back that the customer must be using more energy somewhere else in the house, such as by taking more showers, leaving the back door open, leaving on too many lights, running the pool pump longer, or any number of other actions. That is not to say that the contractor is necessarily wrong . . . but how do you really know?

In Florida, the average mean temperature in 2009 was 4 to 6 degrees warmer than typical, with higher humidity levels and higher overnight temperature levels. This fact, coupled with higher energy costs, caused many customers to question savings they felt they were promised. If your contractor is able to show you the data, and the interpolated savings factors, and you are able to audit this with the knowledge gained from the following factors, you may rest assured that you are reaping the energy benefits and savings you expected.

I promise you that there will be nothing more complicated than simple addition, subtraction, division, and multiplication. (I like to have my 10-year-old help me with this part. Really!)

First, here are the terms and equations you will need to understand:

- *Energy efficiency rating (EER):* Btus of cooling and watts of energy consumption.
- *Seasonal energy efficiency rating (SEER):* Same as EER, but on a national seasonal average. Used for residential air-conditioning systems, typically less than 6 tons capacity.
- *Coefficient of performance (COP):* Kilowatts of heating and kilowatts of energy consumption.

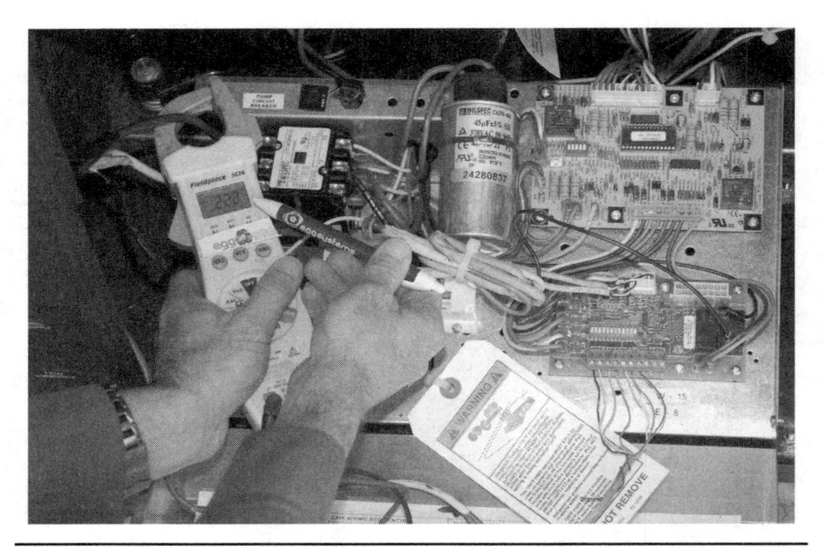

FIGURE 6-2 After installation, be sure to have your geothermal system inspected by a professional. There may be several small adjustments that can be made to ensure maximum energy efficiency, saving hundreds or even thousands of dollars over the lifetime of your system. This chapter will help educate you on the terminology you'll need to make sense of it all. (*Photo by Egg Systems*)

- *kW/ton (kilowatts per ton):* Remember from Chapter 4 that a ton is equal to the removal of 12,000 Btus (i.e., 12 EER = a kW/ton of 1).
- *Heating season performance factor (HSPF):* Btus of heating output and watts of energy consumption. HSPF is a measure of the overall heating efficiency of a heat pump and is similar to an EER rating for cooling. HSPF of 6.8 can be compared with an average COP of 3, and a HSPF in the range of 10 to 11 is marginally accepted in the air-source heat pump world.

Here is a more detailed description of how to find the efficiencies (Fig. 6-2):

Annual Fuel Utilization Efficiency (AFUE) (for Gas Furnaces)

Annual fuel utilization efficiency (AFUE) measures the annual amount of heat actually delivered compared to the amount of fuel supplied to the furnace.

$$\text{AFUE} = 100\, E_o/E_i$$

where E_o = annual fuel energy output (W, Btu/h) and E_i = annual fuel energy input (W, Btu/h).

In commercial chillers, the chiller efficiency depends on the energy consumed. Absorption chillers are rated in fuel consumption per ton of cooling. Electric motor-driven chillers are rated in kilowatts per ton of cooling:

$$\text{kW/ton} = 12/\text{EER}$$

$$\text{kW/ton} = 12/(\text{COP} \times 3.412)$$

$$\text{COP} = \text{EER}/3.412$$

$$\text{COP} = 12/(\text{kW/ton})/3.412$$

$$\text{EER} = 12/\text{kW/ton}$$

$$\text{EER} = \text{COP} \times 3.412$$

If a chiller's efficiency is rated at 1 kW/ton,

$$\text{COP} = 3.5$$

$$\text{EER} = 12$$

Cooling Load in kW/ton

As stated above, the term *kW/ton* is commonly used for larger commercial and industrial air-conditioning, heat pump, and refrigeration systems. The term is defined as the ratio of energy consumption in kilowatts to the rate of heat removal in tons at the rated condition. The lower the kW/ton, the more efficient the system:

$$\text{kW/ton} = P_c / E_r$$

where P_c = energy consumption (kW) and E_r = heat removed per ton.

Coefficient of Performance (COP)

The COP is the basic parameter used to report efficiency of refrigerant-based systems. It is the ratio between useful energy acquired and energy applied and can be expressed as:

$$\text{COP} = E_u/E_a$$

where E_u = useful energy acquired (Btus) and E_a = energy applied (Btus).

COP can be used to define cooling or heating efficiencies for heat pumps:

- Cooling—COP is defined as the ratio of heat removal to energy input to the compressor.
- Heating—COP is defined as the ratio of heat delivered to energy input to the compressor.

COP can be used to define efficiency at single standard or nonstandard rated conditions, or as a weighted average of seasonal conditions. The term may or may not include the energy consumption of auxiliary systems, such as indoor or outdoor fans, chilled water pumps, or cooling tower systems.

IMPORTANT NOTE: Higher COP = more efficient system.

COP can be treated as an efficiency where COP of 2.00 = 200% efficiency. For unitary (packaged) heat pumps, ratings at two standard outdoor temperatures of 47°C and 17°F (8.3°C and –8.3°C, respectively) are typically used.

Energy Efficiency Ratio (EER)

Energy efficiency ratio (EER) is a term generally used to define cooling efficiencies of unitary air-conditioning and heat pump systems. The efficiency is determined at a single rated condition specified by an appropriate equipment standard and is defined as the ratio of net cooling capacity—or heat removed in Btus per hour (Btu/h, sometimes written as Btuh)—to the total input rate of electric energy applied, in *watt hours*. The unit of EER is Btu/watts.

$$\text{EER} = E_c/P_a$$

where E_c = net cooling capacity (Btu/h) and P_a = applied energy (watts).

This efficiency term typically includes the energy requirement of auxiliary systems, such as the indoor and outdoor fans. The higher the EER, the more efficient the system.

Determining Actual Efficiencies

To determine your instantaneous EER rating, have a professional HVAC technician or electrician read the following data for you while the system is operating on full load cooling (set system 3 degrees below actual temperature). For a more complete list of variables, see Fig. 6-3.

Submittal Data—I-P Units

Unit Designation____________________

Job Name____________________

Architect____________________

Engineer____________________

Contractor____________________

Performance Data

Cooling Capacity____________________

EER Calculation: Cooling Capacity___ ÷ Watts___ = EER _______

(If you have an external pump, add watts before division)

EER Full Load_________ EER Part Load ____________

Heating Capacity____________________

COP Calculation: Heating Capacity___ ÷ Watts___ ÷ 3.412 = COP___

COP Full Load _________ COP Part Load____________

Ambient Air Temp____________________

Entering Water Temp (Clg) ____________________

Entering Air Temp (Clg) ____________________

Entering Water Temp (Htg)____________________

Entering Air Temp (Htg)____________________

Air Flow (CFM)____________________

Fan Speed or Motor / RPM/Turns____________________

Operating Weight____________________

Efficiency Calculation Data

Power Supply ____________________

System Amps Part Load (AAC)____________________

System Amps Full Load (AAC)____________________

External Pump Volts Part Load (VAC) ____________________

External Pump Volts Full Load(VAC)____________________

Watts Calculation: VAC____ x AAC____ = Watts__________

Readings must be taken by a certified electrician or HVAC professional

Figure 6-3 Taking accurate measurements in the correct places, you will be able to discern your actual energy efficiency in your situation. This information is valuable in determining whether your geothermal system is running up to its full potential. (*Egg Systems*)

Read the voltage coming into your heat pump and record (V).

Read the amperage of the entire heat pump from the line feeding the main terminal block and record (A).

$$V \times A = \text{W (watts)}$$

If the system is using an external pump (such as a submersible), provide wattage using the method above. Add both wattage figures together for total watts.

Note the tonnage of the system in Btus.

$$\text{Btu/watts} = \text{EER}$$

For example, 48,000 / 2190 = 21.91, or 22 EER.

When the temperature gets within two degrees of set point, it should go into first stage cooling. At that point you can repeat the steps above and determine your part load EER.

For example, 48,000 Btus/1548 watts = 31 EER

In order to determine your instantaneous COP rating (for heating), have a professional HVAC technician or electrician read the following data for you while the system is operating on full load heating (set system three degrees above actual temperature):

Read the voltage coming into your heat pump and record (V).

Read the amperage of the entire heat pump from the line feeding the main terminal block and record (A).

$$V \times A = W \text{ (watts)}$$

If the system is using an external pump (such as a submersible), provide wattage using the method above. Add both wattage figures together for total watts.

Note the heating tonnage of the systems in Btus.

Convert Btus to watts by dividing by 3.413 . . . 48,000/3.413 = 14,064 watts heat output.

220 V × 12.3 A = 2706 W input by method above.

14,064/2706 = 5.20 COP

Determine part load (first stage) COP by allowing the heating set point to come within one degree of actual temperature, and with the system in first-stage heating, take the readings and determine your part load COP.

Here are some of the situations in which you may want to know your actual efficiency:

Often times, by the end of a season of heating or cooling, you will find efficiencies have slipped. This is often due to thermal retention of the loop or boreholes in the summer time, or thermal extraction in the winter. This is not always bad . . . if the fluctuation is acceptable, then you can actually be helping your overall system's efficiency through thermal energy storage as mentioned previously. If these fluctuations tend to be out of a generally acceptable range for most of the season, there may be a problem that can be addressed. See Fig. 6-4 for an example of the fluctuations of temperatures in boreholes.

I have been personally involved in situations in which a geothermal air conditioner has an incoming loop temperature of 100°F or more. This is unacceptable, and we will show you how to guard against being an unwitting victim of this type of problem (see Chap. 9).

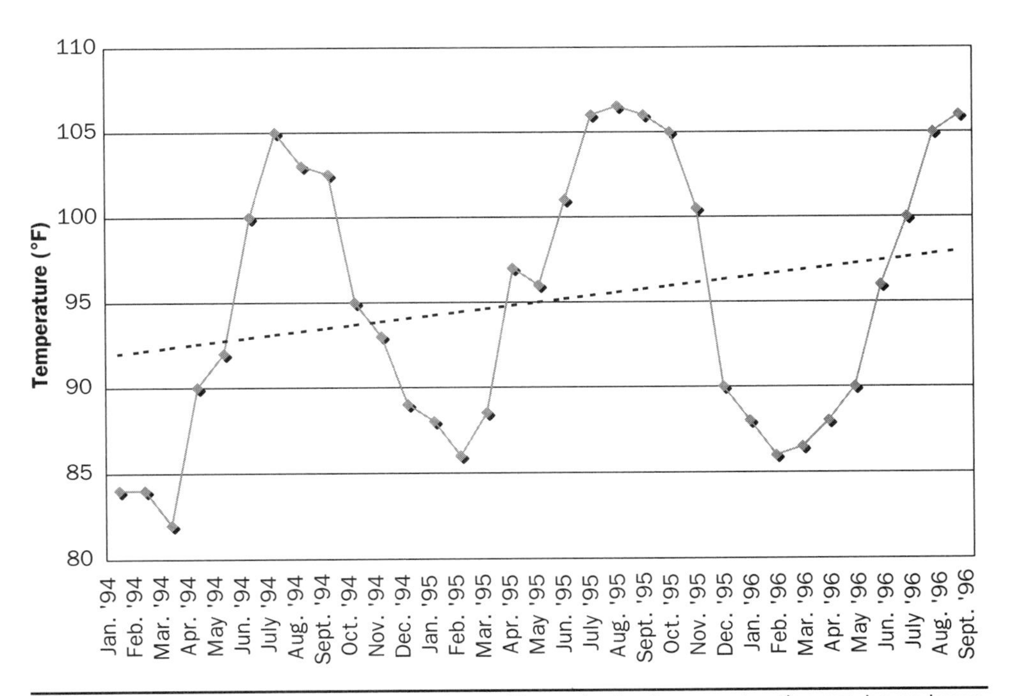

FIGURE 6-4 Borehole field return temperatures. Thermal retention or extraction can be an issue. Certain regions have soil and seasonal characteristics that require departure from commonly accepted practices for loop design. Always consult a professional with plenty of experience. (*John Fontano/Egg Systems, based on a figure by ASHRAE*)

Load Calculations

I remember reading a great *Calvin and Hobbes* cartoon when I was perhaps 10 years old, and it has stuck with me. Calvin and his dad are driving along and approach a bridge weight-limit sign that reads, "Maximum Capacity 10 Tons." Calvin asks, "How do they know the load limit on bridges, dad?" His dad's answer is, "They drive bigger and bigger trucks over the bridge until it breaks. Then they weigh the last truck and rebuild the bridge."

Load calculations are a lot like that...if you don't do them properly.

Back in Chapter 4, I talked about the difficulty with which I did my first load calculation on that 10,000-foot house in Carrollwood. That was 1990. Now in 2010, it's become infinitely more difficult. Not only are there heat gain and loss analysis forms that must be completed, but there are also energy calculations required in many states and jurisdictions.

Energy calculations take into consideration everything within the building that you can imagine would have anything to do with heat gain or loss. This includes the basics, such as the air conditioner, hot water tank, windows, insulation, and color of the roof. But it also includes the types and watts per square foot of lighting. This must all be completed by the architect or engineer, and these professionals often expect the HVAC contractor to do this calculation, since the mechanical system is the single greatest source of both energy consumption and energy distribution in most cases.

So, on top of needing to be aware of the most complex systems in a building, we in the field of HVAC need to calculate all of the heat gains associated with each of the other trades. That means we have to interview and audit each of the other subs and the general contractors to get an accurate energy calculation. The computer programs alone can cost thousands of dollars, and the experience, skill, and time factors to do these calculations are highly prohibitive to smaller firms.

I have had the privilege of putting many highly skilled employees into business for themselves. Some of my best training has grown wings and flown. But I'm not bitter, OK? Seriously, there are many associated contractors, whether former employees, just good friends or both, that share work with my company. We provide these services to many other good contractors who wisely choose not to make the investment at this time. My point in telling this story is that mechanical calculations for sizing of systems and for energy analysis are difficult at best. I could not have made it through the process at first without a lot of help. It's much more difficult now in this age of energy efficiency awareness. Don't

FIGURE 6-5 Pond or lake loops are common for homes along waterways, since they are so easy to install. However, there are also many situations even along a waterway in which the better system may involve a ground loop or pump and reinjection. (*Photo by Egg Systems*)

trust the sizing of your ducts, air-conditioning, and geothermal source to someone without proper credentials (see Fig. 6.5). You should feel free to ask for a copy of the appropriate paperwork.

Manuals Published by the Air Conditioning Contractors of America

There are certain terms that you should know in order to speak the language of heat gain analysis and energy calculations. The Air Conditioning Contractors of America (ACCA) has provided many of the accepted guidelines, along with the American Society of Heating, Refrigerating, and Air-Conditioning Engineers, Inc. (ASHRAE). These include the following:

Manual J: Residential Heat Gain and Loss Analysis

It's common language to say, "Have you performed a manual 'J' on this job?" That is the same as asking if your contractor has performed a building-specific heat gain and loss analysis. If he or she has, they'll have a copy of it in their files. You're paying a lot of money to have a new system put in your home. In Chapter 9, we'll go over what your contractor's proposal should look like.

Manual N: Commercial Heat Gain and Loss Analysis

Commercial heat gains differ from residential because of different hours of operation and usage patterns. For example, on a 45°F February day, I may need to run the air-conditioning in my office, while I run the heat at home. At the office, I'm surrounded by copiers, computers, printers, lots of lights, and few windows. You will see similar differences in a high school gym, cafeteria, warehouse, factory, or church. Suffice it to say that loads vary significantly.

Manual D: Residential Duct Design

When you ask if your duct sizing is sufficient, it's always helpful to note if you were having any balance or supply issues up to this point. Often times, heating-only duct systems are not sufficient to provide cooling; there is simply more cubic feet per minute (CFM) of air required to cool a building properly under most conditions (and this is why it can be especially expensive to retrofit older homes with central air). But note that when properly installed, and with adequate ducts, your heat pump system will automatically adjust your fan speed to accommodate proper airflow. Still, it's not a bad idea to request a printout that shows the CFM and duct size of each room you are cooling and heating.

Manual Q: Commercial Low-Pressure Duct Design

There are situations where you need different designs for ductwork in commercial applications. Standard ventilation needs can often be handled with low-pressure systems, while larger buildings, such as arenas or high-rises, may require high-pressure duct systems. There may be fan coils and variable air volume dampers to allow adjustments to specific sections of a building. Also, you may encounter fan-powered controllers, barometric relief dampers, exhaust fans, and 100% fresh-air systems, for rapid air turnover. The important thing in commercial duct design is to ensure that you are using a competent contractor and engineer. Check references!

Energy Calculations and Value

Energy codes have been passed by many local, state, and national governments around the world to protect consumers and the environment.[1] Typically, they mandate a minimum efficiency standard per square foot, although they can be quite detailed.

When it comes to energy-saving equipment, you will often hear different options, with costs ranging from reasonable to pricey. The point of diminishing returns can only be settled upon by you and your

[1] In the United States, the Web site www.energycodes.gov is a great place to start to make sure you are in compliance. At Egg Mechanical, we comply with the stringent guidelines of ASHRAE 90.1-2007/2009 IECC.

budget, as well as your patience. You should also give consideration to life cycle and longevity values (Fig. 6-6), which will be covered in greater detail in Chapter 13.

Remember, if you do not ask your contractor for comparisons, you may not get them. What you will probably get will be a capable system, but one that costs more than you need to pay. So, simply ask the question: "How much does this item save in my energy costs versus the cost of the item to me?"

I was recently in a large office-building meeting with a facilities manager. They were considering a geothermal system, and the pitch seemed to be going well. Another individual had invited a supplier's representative, who was speaking about the performance merits of the earth-coupled heat exchanger he represented. After all of the promises made by this rep, the facilities manager asked him for some references

FIGURE 6-6 This WaterFurnace geothermal air-conditioning system installed with the lake loop in the previous figure looks like it did the day it was put in nearly 20 years ago. (*Photo by Egg Systems*)

for the product. . . . The room was silent. The rep looked stunned, as if no one had the right to do such a thing. He didn't actually respond at that time and the awkward silence was broken by a change of subject. After the meeting, as I walked to my truck with this gentleman, I made it clear that I would not put my reputation on the line for this product. I would use it in a test with a redundant system and only with fully acknowledged disclosure from the customer.

Don't test unless you know that everyone is on board and prepared for problems. It's expensive and can be disastrous. Otherwise, don't start anything without complete data.

Summary

Having learned how to calculate system efficiencies and heat gain and loss, give another thought to the story of the bridge weight capacity. If you have received a proposal without the proper documentation of these details, then skip to Chapter 9, where we mention how easy it is to place a restraining order on the contractor who is bothering you. It may sound funny, but unfortunately it happens far too often.

CHAPTER 7

Understanding Pricing of Geothermal Systems

When I was 14, I was told to go into construction by a kindly old gentleman who was sitting on the sofa in the foyer at church. I still remember the orange cushion and the dark brown wooden legs ...on the sofa, not the old guy. I asked him, why construction, and he answered, "You'll get rich." He said I could charge any price and people would pay it.

Even at 14, I was skeptical. I wondered why I could charge anything; wouldn't there be competition? He said there is "no competition," because everyone who's able to work is busy, and you almost have to pay them to come out and even look at your project. So I asked him how someone could get involved. He said just walk on a site, pick up a hammer, and go to work. Of course, that didn't work ... at least not in my situation.

My father and grandfather were skilled in most facets of construction, but primarily in plumbing and electrical work. With the exception of a few drywall and framing jobs I did with my younger brother on the side, I pretty much stuck to the mechanical and electrical trades. But what the old guy on the sofa said really stuck in my mind over the years. Not so much the get rich part, but the way he indicated that it would happen: "You could charge any price and people would pay it." I made a resolution early on that greediness is not the answer, and that if one does get greedy, he won't make it very long in business. Greedy or crooked contractors can often be seen on TV in consumer alerts. "Channel 10 exposes crooked plumbers in Augusta....Story at 11." Unfortunately, these reports probably only catch a small fraction of those problems.

The best defense is an informed consumer. The next best defense is a well-informed contractor. Between the two, there is mutual respect and communication that exists in good measure. The latter is the primary purpose of this book. It is to inform the consumer and contractor alike what to expect from conception throughout the process of installation, and on into the commissioning and proving of the geothermal system. This, as you should be aware by now, is not

FIGURE 7-1 With a few exceptions, if you are getting a geothermal system, you will have some heavy equipment in your yard for a couple of days. This drill rig was in a Georgia yard for 2 days, drilling vertical loops for a 4-ton system. (*Photo by Egg Systems*)

a technical textbook on how to install geothermal air-conditioning systems, but rather a book on how to understand the necessary terminology, steps, and measures to protect the consumer and this industry from ignorance and misinformation.

We conducted interviews and received answers to hundreds of surveys sent to qualified contractors throughout the world. Ted Chittem of Air Perfect, Inc., in Milford, Connecticut, warned us via phone that, "Buying a heating and cooling system is not like buying a fridge or TV. When you do your research, take it out of the box and plug it in; it is not plug and play, and the reality of the overall project is absolutely going to be tied to choice of equipment and execution of the installation." (Fig. 7-1)

Let us quote from another response that came over e-mail:

QUESTION: What problems have you seen with each type of geothermal HVAC system?

ANSWER: "Poor design and very poor installation. Homerun shots too small. Pump stations under- and oversized. Ponds too small. Contractors using pond water to fill a closed loop and sucking in mud."

QUESTION: Are you worried about poor quality of contractors entering the market?

ANSWER: "Yes."

QUESTION: What stories do you have of unqualified contractors?

ANSWER: "Too many to write about."

QUESTION: Are your distributors making sure newcomers are certified somehow? What is that certification standard?

ANSWER: "Manufactures are doing a horrible job. They need to inspect a portion of jobs installed, and a much higher percentage of a newcomer's first jobs."

We can assure you that standards have been developed, and are in the process of being enforced. But there's more:

QUESTION: What is the biggest problem you are facing right now?

ANSWER: "Financing for potential purchasers."

QUESTION: Biggest blessing?

ANSWER: "The investment it takes for contractors to get into this technology."

Basically, Alan Givens with Parrish Services in Virginia, the contractor quoted above, is saying that most contractors are staying out because it's too expensive to be trained and outfitted properly. There is also a problem with financing for consumers, which will be addressed in Chapter 11. I just left a training meeting Egg Geothermal conducted in a bank's boardroom, wherein we proved that "net positive cashflow" financing could be easily attained with today's rebates and geothermal technology. We've had several such meetings in recent months with banks, in which we train loan officers on how to help their customers take advantage of tax credits and incentives available for geothermal. These usually go very well.

IMPORTANT NOTE: *As a general rule, you should spend no more than $9000 per ton for geothermal systems, including a loop and everything you need to install it in a new home.*

There are different margins for different contractors in different areas. Plus, there are differing qualities and efficiencies for equipment. This price is basically a starting point for a top-of-the-line, name brand, 30 EER system. A standard duct system will be up to $2000 of that per-ton cost.

There are certain types of geothermal systems where this cost can be significantly reduced. Examples are pond loops and larger pump and inject systems, or systems where geothermal sources are readily available, such as in the case of existing water wells. This is primarily due to the nature of the costs of excavation. A pond loop will require more pipes, or a larger exchanger, but generally, the labor does not increase at the same rate. The same is true with a well water system. A well that works for a 3-ton system can often work for a 10-ton combination of systems without any additional costs except for the pump size. Though the cost of a larger pump will be higher, the drilling costs will not often have to increase.

Factors That Affect the Price of Geothermal HVAC

There are several factors that can cause a variance from the pricing above, including upgrades, downgrades, efficiency ratings, quality, topography, heavy sales volume, and competition. In Figs. 7-2 and 7-3 you will see what differing loops cost per ton, as well as the pricing for different equipment efficiencies and quality standards. It is important to note that much of the equipment used to this point in the industry is of good quality, easily lasting 25 or 30 years with normal operation. I have personally installed many systems that are now 20 years old and have had no problems with quality of equipment installed from the three major manufacturers that I have used.

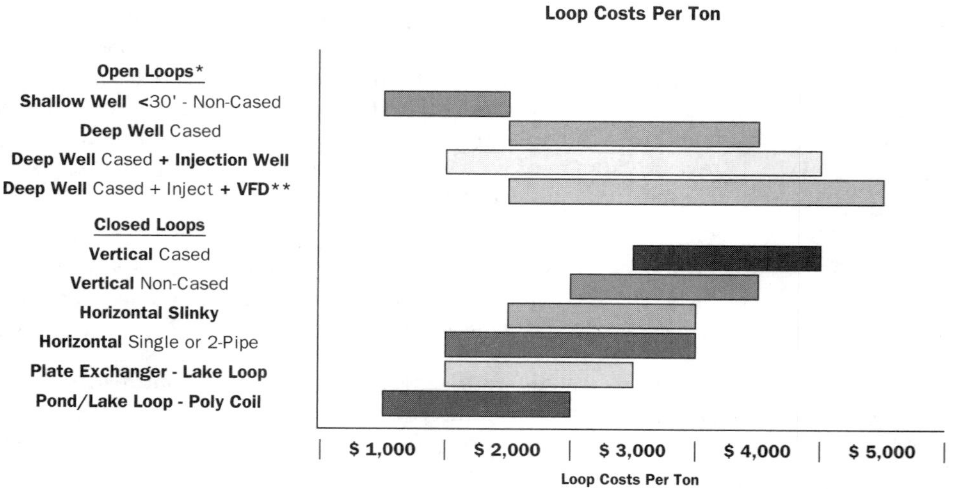

*In Open Loops the cost per ton reduces significantly as the system size increases.
The cost per ton will also vary depending on the drilling depth required for a quality water source.
**VFD - Variable Frequency Drive. The VFD is a high efficiency pump for the supply well.
It provides only the amount of water being called for at any given time, and consumes less energy than a standard pump operating at a constant speed.

FIGURE 7-2 Retail cost for equipment. (*John Fontano/Egg Systems*)

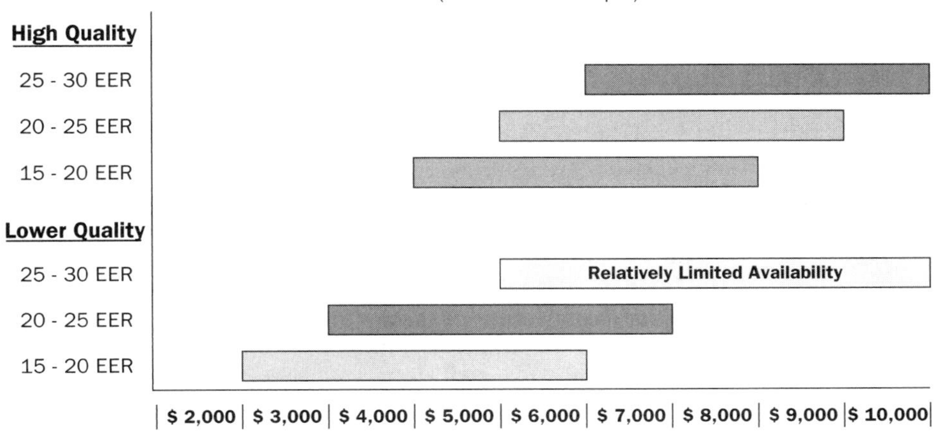

FIGURE 7-3 Loop costs per ton. (*John Fontano/Egg Systems*)

Efficiency Ratings

Efficiency ratings are the essence of energy savings. They are most likely the reason that you are reading this book. The higher the efficiency rating on a system, the more money and energy you will save each day, month, and year. This is where it is important, however, to look at your payback analysis, performed by a competent engineering company. You will usually find that there is a point of diminishing returns, wherein you can make the decision to settle on the best bang for your buck.

Remember, efficiency ratings are typically presented as EER or COP, with EERs normally in the 15 to 30 range, and COPs in the 3 to 6 range. To see what the difference is and how to calculate these values, flip back to Chapter 6.

Quality

Right now, the quality of geothermal products is good among the brand-name manufacturers. However, one concern we face is the cycle of planned obsolescence. One geothermal air-conditioning system my company maintains in a downtown Tampa church is chugging along on a 1938 reciprocating, water-cooled compressor. I remember thinking about 10 years ago that the system probably didn't have much time left, back when Egg Geothermal converted the old R-12 refrigerant compressor to the more environmentally friendly R-409A. Just a few months after that it began to lose refrigerant, and I felt that I was going to be right. But a simple retightening of the bolts that hold the head gaskets in place fixed the problem, and it's been running fine ever since. Seventy-two years is not too bad for longevity!

On the other side of the spectrum, I think of residential and commercial air-conditioning equipment that routinely fails right at 9 to 12 years of age. The adage that "they don't make them like they used to" is sadly true for many industries. So with regard to quality, all that I can say is to keep an eye on consumer reports you trust. The products that rank well now, such as ClimateMaster, WaterFurnace, and Florida Heat Pump, are probably going to be a good buy. We hope that we continue to see the geothermal heat pump industry maintain high equipment standards. But in reality, we will likely see inferior equipment slip into the market. As this happens, we will see equipment failures in the first years and months, and we will have problems getting replacement parts. Much like a friend on my street who has a no-name four-wheel ATV that needs a rear end part . . . he can't find it anywhere at any price. Well-built products do cost more, but they are very much worth the price.

System Sizes

As the tonnage of systems increase for residential and commercial projects, you will often see a decrease in the cost per ton. I've reviewed payback analyses that are at 5 years on 3- and 4-ton equipment, and I've reviewed paybacks of just months or even negative factors on projects with, say, five systems and 20 tons of air-conditioning.

Topography

Perhaps not surprisingly, the lay of the land can impact the cost of system installation. If the soil is rocky, as in New England, it can necessitate vertical loops, which generally cost more than horizontal systems. Similarly, the orientation and plan of your building will affect the HVAC design (Fig. 7-4).

Load Sharing

This was covered in some detail early in Chapter 3. There is a premium cost associated with setting up load sharing, but this begins to get tricky. . . . You need to find out what is eligible. For example,

FIGURE 7-4 Geothermal air-conditioning systems can be installed in most any yard. The silhouette of the drill rig in this tropical Florida neighborhood is almost lost against the trees and behind the Egg Geothermal work van. (*Photo by Egg Systems*)

swimming pool heat pumps are not eligible under U.S. federal tax credits for energy efficiency, but they are highly advantageous for load-sharing applications. An engineered drawing must be completed for a load-sharing circuit, indicating the piping, valves, controls, and gallons per minute to complete the system. There is no way that load-sharing systems can be roughly thrown together and operate properly.

IMPORTANT NOTE: *Always make sure you enlist the services of a professional engineering firm to design your systems.*

Load sharing is easily the best efficiency upgrade to an already well-designed geothermal HVAC system. Let's say that your equipment is operating in the cooling mode. The byproduct is waste heat, conveniently contained in a plastic pipe moving along the condenser water piping. With properly piped load-sharing circuits, that byproduct is sent through the hot-water-dedicated heat pump, utilizing nearly all of the heat, and increasing efficiency dramatically. In a large residence or commercial building, you can take advantage of load sharing to a degree that is truly hard to believe. We need logarithms to figure it all out. But that's for our engineers to take care of. Just sit back and prepare to be amazed.

Sales Volume and Competition

Heavy volume and competition can be a factor in any industry. These factors should bring down the cost of manufacturing and installation of products, although sometimes heavy volume can result in delays, as well as upward pressure on pricing. (Remember the older gentlemen who told me I could charge as much as I wanted, because it's so hard to get contractors to come out and look at your project?)

Egg Mechanical installed thousands of HVAC systems for new homes and shopping centers during the last decade (Fig. 7-5). At one point, a good friend of mine from church, Sam Harbert, asked me to work on some tract homes for the national builder US Homes (a division of Lennar since a 2000 buyout). Sam told me the amount the company was willing to spend on HVAC systems, but it didn't seem possible for me to get down to that number. He convinced me to give it a try, and I lost money for about the first six months, on about 100 homes. Then, I began to break even. Eventually, I started to turn a profit of a few hundred dollars on each new home. I would not have believed it early in my career that I would do $9000 and $10,000 tract home air-conditioning jobs for just a $300 or $400 profit. But we did well, and I felt good about giving my customers quality jobs at low prices.

Optional Upgrades

Much like a car, there are several options that can be added to a geothermal air conditioner. Let's take a look at the most common ones.

Heat Recovery for Domestic Hot Water

One common option is a hot water generator that can supply much of the domestic hot water needed for a home or business. Higher efficiency systems that are designed for air-conditioning usually can't provide hot enough water for all of a building's needs, and will need an additional holding tank and electric or gas boost. The trade-off for higher efficiency is well worth the extra cost for hot water. Remember that you can also choose a dedicated domestic hot water heat pump, mentioned in Chapter 3, and explained below.

Until fairly recently, hot water recovery systems were required for geothermal units to receive tax credits. Some installers are glad this rule was loosened, because they argue that the equipment doesn't always make sense. "I typically don't recommend [hot water generators], because relative to their cost they're not worth it," Ted Chittem told us.

FIGURE 7-5 A reputable and experienced company will make your geothermal air-conditioning system installation go smoothly. Preplanning is the key. Every detail will be addressed, right down to replanting the flowers in the front yard. (*Photo by Egg Systems*)

Domestic Hot Water Geothermal Heat Pumps

In many cases, a dedicated domestic hot water geothermal heat pump can be a good buy. Unlike heat recovery units that produce heated domestic water while the air-conditioning system is cooling or heating, these heat pumps work on demand to fill a tank suited and sized for any application. They are typically the most efficient and cost-effective way to produce hot water, besides solar thermal systems. But solar is sometimes not available when we need it the most. So if your family is like mine, you can take that shower at 9 p.m. or 5 a.m. with a little less guilt. Or you may feel free to start a load in the washing machine and the dishwasher.

I know that my wife and children were getting a little tired of my "solar timing" for use of hot water. I think it was the sweet way that my wife referred to me as either the "greenest"...or was that the "meanest" man alive that cold night last November? Either way, we're grateful for the new dedicated geothermal heat pump water heater. No more worrying about the energy consumption at night!

Exchanger Materials

One potential upgrade to geothermal systems is in the exchanger materials. The standard is usually copper, with upgrades to cupronickel and titanium available, often for a price of hundreds or even thousands of dollars. Nickel and titanium alloys are more resistant to any corrosive chemicals that may be present, such as sulfur, iron, or manganese, as well as fouling agents, which can cause mineral or scale build up, reducing flow and impeding heat transfer. If you are going with an open loop, it's important to test the water that will be used for exchange, to see what minerals and contaminants may be present.

Sometimes debris that is abrasive, such as sand, may be in the liquid. I have seen instances wherein stainless steel pump impellers have been ground to nothing in just a few years by the presence of a little sand in a system.

Compressor Stages

Compressors can be single- or two-stage in residential applications. Commercial variations are too numerous to mention. A single-stage compressor system always runs at 100% capacity, while a two-stage runs at 60% or so most of the time, giving a much better efficiency rating, as well as other benefits such as better humidity and temperature control.

A good way to look at this feature is to compare it once again to an automobile. A single-speed system is much the same as a car that has only two speeds: full throttle or off. Can you imagine how bad the fuel economy would be? I remember my dad lecturing me about the loss of fuel economy resulting from stomping on the gas.

On commercial equipment, we can take advantage of an increasingly common piece of technology, the *variable frequency drive* (VFD). Basically, a VFD gives motors the ability to change their cycle speed, essentially giving them a smooth acceleration and cruise control. Otherwise, motors are typically stuck at 60 Hz (cycles per second), due to the standard for grid power (at least in the United States). You can be certain that a similarly effective system will be coming to home geothermal HVAC in the future, at a reasonable cost.

Hot Gas Reheat

Another upgrade available on some models is *hot gas reheat*, which is typically expensive (in the thousands of dollars). This allows precise humidity control in the toughest of situations with the highest efficiency possible. We should clarify that the "hot gas" used is not a fossil fuel. The term actually refers to the state of the refrigerant gas in the refrigerant cycle just after it is compressed and exits the compressor. The gas is rerouted—when humidity controls call for dehumidification—to an additional refrigerant coil placed just down the air stream from the evaporator, or cooling coil.

If you're confused, let us try to explain in layman's terms. Every time air passes over the evaporator coil (the cold coil, 40 to 50 degrees typically), the water or humidity in the air condenses on the fins and falls into a condensate pan (see Fig. 7-6). It's not always possible to get out all of the humidity necessary for optimum comfort on the first pass of air...and if you go through the coil too many times the air temperature will get below a comfortable level. This will often give you

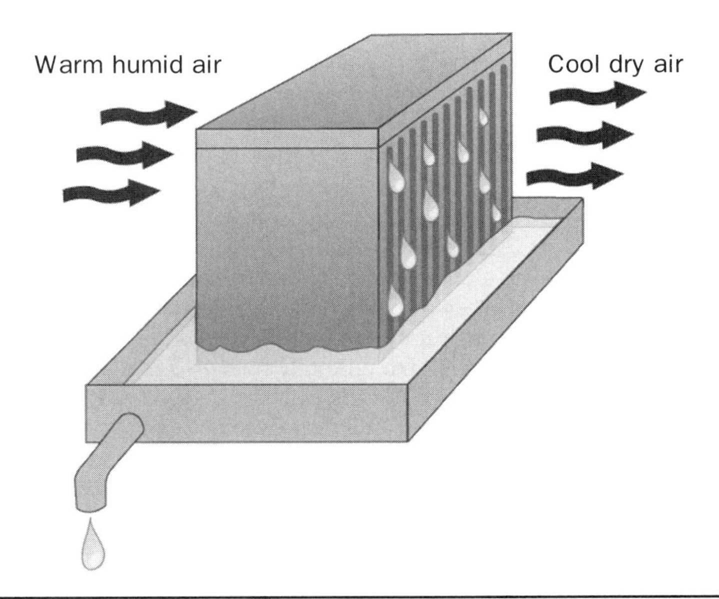

FIGURE 7-6 A key to comfort is controlling the humidity level in the home or business. Geothermal air-conditioning systems excel at dehumidification as well as energy efficiency. (*Sarah Cheney/Egg Systems*)

the sensation of feeling cold and clammy. So, very simply, if the air is reheated at a prescribed amount for a long enough period of time, the perfect humidity level can be achieved.

Many commercial buildings use reheat of some kind to precisely control humidity. But here's the unfortunate part: many buildings use a separate heat source, such as electricity or gas to reheat the air, increasing energy consumption and, in many cases, burning fossil fuels.

Intermediate Exchangers

The first time I saw a pair of blue nitrile gloves was about 10 years ago at the Ford dealer that was servicing my truck. I had used latex gloves for messy service projects, but I was fascinated by these tough gloves, and talked to the mechanic about their effectiveness. He said, "They keep my hands clean and free from cuts and pinches." While I prided myself on the calloused hands that I had developed early on in my life, I generally like to protect them as I get older.

Intermediate exchangers are like a good pair of gloves that can be cleaned after heavy usage. An intermediate exchanger goes in the circuit between a loop or a well, and circulates water through the equipment. It also aids in the ability to more effectively load share. The cost of an intermediate exchanger can range from a couple thousand dollars and up. It will be placed outside normally and will provide a single point for easy cleaning on a periodic basis.

Direct Digital Controls (DDC)

Direct digital controls (DDCs) allow you to treat your home or building like a piece of computer hardware. If you get a new gadget for your computer, such as a wireless keyboard, you can plug it in and it works, even if it is made by a different manufacturer. That's like DDC; all of the systems of your home can be networked, so you can control them with smart programs that help them to share energy with one another, and allow you to remotely turn on lights, unlock doors, water the grass, run the dryer and dishwasher, and so on. Plus, get instant efficiency ratings. The system will be able to anticipate problems and send you warnings, or automatically call service technicians.

Much like a spell check program, the systems of the future may prompt, "Do you really want to start the washing machine now, while the shower is on and the dishwasher is running? Please allow me [the smart DDC system] to start it for you later, when energy and water consumption can be optimized, saving you 19.4 cents." This would allow your geothermal domestic hot water heat pump to run longer at a COP of 5, without the use of the emergency back-up heating element, which has a COP of less than 1. Plus, it will enable you to use less energy during peak periods, when electric rates may be higher. It's all coming, and quicker than we think.

When we all connect to the *smart grid* in the coming years, the enhanced electronic component of the power grid will allow seamless integration with home DDC controls. You'll be able to monitor and operate everything from any Internet connection.

IMPORTANT NOTE: *Know what you're getting into.*

Summary

Most of the long-timers in geothermal air-conditioning are in this for the duration, and we want to give our customers the best system for the lowest price we can. My sales managers and CFO regularly chastise me for under quoting. During an interview with John Bailey of ClimateMaster, he and I chuckled that Dan Ellis, their president, is much the same way. He'd give it all away if he could. He has a passion for the technology that comes out in all that he does and says. And the millions of dollars that have been spent by ClimateMaster in passing legislation to get this technology off the ground are a testimony to that. I mused with that thought as I reflected on the $823,000 spent in my marketing efforts in Florida and Georgia in just under a year.

My point is that as geothermal air-conditioning gains popularity and becomes mainstream, prices will stabilize, although there will still be a number of factors that affect cost, such as options, equipment quality, and the local market. We'll see more efficient and versatile equipment, and a higher degree of competence from the average contractor. Remember: Always get an engineered system!

CHAPTER 8

Incentives, Tax Credits, and Rebates

Incentives are an interesting thing. As I travel around to meet with wealthy homeowners and business owners, regular hard working people, and employees or board members for properties that may utilize systems, I seem to hear one overriding theme. The conversation usually goes like this:

> "Are you aware of the significant federal tax credits and incentives available for your home and business if you take advantage of geothermal cooling and heating technologies?"
>
> "Yes, and I think it's a shame that we're spending our tax dollars so recklessly on people's homes and businesses. The government should just let things take care of themselves.... But if it's there, I'm going to take advantage of it....I'd be a fool not to!"

Often times, depending on the conviction and persuasion of their political views, the conversation can get rather colorful. However, I can say that I have come very close to perfecting a type of neutrality. Much of this stems from the fact that my language is pretty straight forward... no curse words, only a wince or two. There is a saying that my family tries to live by when considering another person's views: "Always assume noble intent."

My father was a firm believer in incentives. He had a way of encouraging assignments by offering a nominal reward, such as a small allowance, but there would often be a bigger reward for an effort that went above and beyond. For example, I remember finding a five-dollar bill in the shop he had asked me to clean out. I never would have found it had I not gotten down on my belly and removed the debris from underneath a workbench with a broomstick. But there was the prize, with the affixed note: "Valiant effort, job well done!" To this day, I enjoy lessons like this with my own children. I expect the finest efforts and reward them well for it, which happens more often than not.

FIGURE 8-1 While trying to negotiate tax credits may seem trickier than installing HDPE piping, it is actually easier than you might think. (*Photo by Egg Systems*)

Still, it's easy to believe that were it not for minimum standards, we would still be using lead-based paint and asbestos insulation. Unfortunately, many of us can't see past the initial cost, even if we have noble intent. So a powerful way to get progressive technologies (Fig. 8-1) into the mainstream is to provide some type of incentive or initiative, at least until the sector matures and gets a solid footing. Plus, in many cases, upstarts must compete with the established tax breaks and lobbying power of well-heeled legacy businesses. This is certainly true when it comes to anything related to fossil fuels.

Egg Systems has installed air-conditioning systems in thousands of new homes and businesses. Less than one in a hundred elect to purchase anything more than the minimum standard required by law. To me, that suggests that many people would opt for a lower standard if it were available, even when the payback is shown to be less than 10 years. On a mortgage, that should be an easy choice. This is especially true when many have added an additional $14,000 in options to the master bathroom alone, but not an extra $1000 for a higher efficiency HVAC system. It was only with the greatest of self-discipline that I elected to have my plumber install a solar thermal panel for the addition to our home a couple of years ago. I am so very grateful that I did it now. But that extra $5000 seemed almost too much at the time.

So here we are with energy efficiency incentives in the form of tax credits. But that's not all that is available. Allow us to help you see the full range of possible support. We will sprinkle in a healthy portion of what is planned to come . . . and what impact it may have on us.

U.S. Federal Tax Credits

In November of 2008, Egg Geothermal installed a system for a customer in the Tampa Bay area. This wasn't much different from the hundreds of other installations we've done over the years, with one exception.

My project manager asked me where he could get a federal geothermal tax credit form. I remember laughing to myself and thinking, "Oh, what poor misguided soul thought there was any hope of ever seeing federal tax credits for geothermal air-conditioning?" When he persisted that the customer knew of something, I asked him to research the subject further. The next day, he came to me with a small stack of papers, with a heading that read, "American Recovery and Reinvestment Act of 2009." Within the pages of this summary, we found that geothermal heat pumps were eligible for a 30% uncapped tax credit.

I immediately commissioned our Web designer to develop a geothermal HVAC-focused Web site, and began to get the word out about this new opportunity. It felt like some of my zeal for the technology was starting to spread.

In the early 1990s I had to wrestle with the utilities for rebates, beg for help from local government, and try to educate local builders. On more than one occasion, I wished that I could have had the time back that I had believed I foolishly wasted. It is wonderful that earth-coupled air-conditioning has finally received recognition from the federal government. There is so much training and educating to do, but I'm grateful to be a small part of that.

As of December 1, 2009, homeowners who install geothermal heat pumps with Energy Star certification are eligible for a 30% federal tax credit. Note that, on average, Energy Star-qualified products will be at least 45% more energy efficient than standard options (meaning the best of standard air source systems in this case). New requirements for water-to-air and DX models will take effect on January 1, 2011. Even more stringent levels will go into effect for water-to-water and water-to-air models on January 1, 2012 (see Fig. 8-2).

To take advantage of the 30% tax credit, simply purchase a qualifying geothermal air conditioner. After completing the job, your contractor will provide you with a sales form that you can keep handy in your files in case of an audit. You will then need to procure an IRS form 5695 (Fig. 8-3) for the year you are applying for the tax credit. This is where it's a good idea to consult your tax professional, but the

Geothermal Heat Pump
Energy Star 3.0 Specification

Table 1: Tier 1 Requirements (December 1, 2009)		
Product Type	EER	COP
Water-to-Air		
Closed Loop Water-to-Air	14.1	3.3
Open Loop Water-to-Air	16.2	3.6
Water-to-Water		
Closed Loop Water-to-Water	15.1	3.0
Open Loop Water-to-Water	19.1	3.4
DGX		
DGX	15.0	3.5

Table 2: Tier 2 Requirements (January 1, 2011)		
Product Type	EER	COP
Water-to-Air		
Closed Loop Water-to-Air	16.1	3.5
Open Loop Water-to-Air	18.2	3.8
Water-to-Water		
Closed Loop Water-to-Water	15.1	3.0
Open Loop Water-to-Water	19.1	3.4
DGX		
DGX	16.0	3.6

Table 3: Tier 3 Requirements (January 1, 2012)		
Product Type	EER	COP
Water-to-Air		
Closed Loop Water-to-Air	17.1	3.6
Open Loop Water-to-Air	21.1	4.1
Water-to-Water		
Closed Loop Water-to-Water	16.1	3.1
Open Loop Water-to-Water	20.1	3.5
DGX		
DGX	16.0	3.6

FIGURE 8-2 The efficiency of geothermal air-conditioning systems will increase according to Energy Star requirements for the next several years. (*ClimateMaster*)

gist of the credit is straightforward. This is not a tax write-off, but a tax credit. So if you owe $10,000 in federal income tax, and you have a $12,000 credit due to you because you bought a $40,000 system, you will get $10,000 back. And then next year, you will get the other $2000 back. The bad news to those who do not pay taxes for any reason is that the IRS is not going to print money just to give to you. They will only give you credit for taxes you were going to pay anyway.

Before we depart from residential tax credits, there are a few more criteria to keep in mind. According to the Database of State Incentives for Renewables and Efficiency[1] Web site:

> A taxpayer may claim a credit of 30% of qualified expenditures for a system that serves as a dwelling unit located in the United States and is used as a residence by the taxpayer. Expenditures with respect to the equipment are treated as made when the installation is completed. If the installation is on a new home, the "placed in service" date is the date of occupancy by the homeowner. Expenditures

[1] DSIRE, www.dsireusa.org, is an excellent resource for finding region-specific incentives for a wide range of renewable and efficiency technologies.

Form **5695**

Department of the Treasury
Internal Revenue Service

Residential Energy Credits

▶ See instructions.
▶ Attach to Form 1040 or Form 1040NR.

OMB No. 1545-0074
2009
Attachment Sequence No. **158**

Name(s) shown on return | Your social security number

Before You Begin Part I: Figure the amount of any credit for the elderly or the disabled you are claiming.

Part I **Nonbusiness Energy Property Credit** (See instructions before completing this part.)

1	Were the qualified energy efficiency improvements or residential energy property costs for your main home located in the United States? (see instructions) ▶	1	☐ Yes ☐ No
	Caution: *If you checked the "No" box, you cannot claim the nonbusiness energy property credit. Do not complete Part I.*		
2	Qualified energy efficiency improvements (see instructions).		
a	Insulation material or system specifically and primarily designed to reduce the heat loss or gain of your home	2a	
b	Exterior windows (including certain storm windows) and skylights	2b	
c	Exterior doors (including certain storm doors)	2c	
d	Metal roof with appropriate pigmented coatings, or asphalt roof with appropriate cooling granules, that are specifically and primarily designed to reduce the heat gain of your home, and the roof meets or exceeds the Energy Star program requirements in effect at the time of purchase or installation	2d	
3	Residential energy property costs (see instructions).		
a	Energy-efficient building property	3a	
b	Qualified natural gas, propane, or oil furnace or hot water boiler	3b	
c	Advanced main air circulating fan used in a natural gas, propane, or oil furnace	3c	
4	Add lines 2a through 3c	4	
5	Multiply line 4 by 30% (.30)	5	
6	Maximum credit amount. (If you jointly occupied the home, see instructions)	6	$1,500
7	Enter the smaller amount of line 5 or line 6	7	
8	Enter the amount from Form 1040, line 46, or Form 1040NR, line 43 — 8		
9	Enter the total, if any, of your credits from Form 1040, lines 47 through 50, and Schedule R, line 24; or Form 1040NR, lines 44 through 46 — 9		
10	Subtract line 9 from line 8. If zero or less, **stop.** You cannot take the nonbusiness energy property credit	10	
11	**Nonbusiness energy property credit.** Enter the smaller of line 7 or line 10	11	

For Paperwork Reduction Act Notice, see instructions. Cat. No. 13540P Form **5695** (2009)

FIGURE 8-3 IRS form 5695 is required to apply for your U.S. income tax credit, which is 30% of the cost of your geothermal air-conditioning system. (*IRS*)

Form 5695 (2009) Page 2

Before You Begin Part II:

Figure the amount of any of the following credits you are claiming.

- Credit for the elderly or the disabled.
- District of Columbia first-time homebuyer credit.
- Alternative motor vehicle credit.
- Qualified plug-in electric vehicle credit.
- Qualified plug-in electric drive motor vehicle credit.

Part II **Residential Energy Efficient Property Credit** (See instructions before completing this part.)

Note. *Skip lines 12 through 21 if you only have a* ***credit carryforward from 2008.***

12	Qualified solar electric property costs		12
13	Qualified solar water heating property costs		13
14	Qualified small wind energy property costs		14
15	Qualified geothermal heat pump property costs		15
16	Add lines 12 through 15		16
17	Multiply line 16 by 30% (.30)		17
18	Qualified fuel cell property costs	18	
19	Multiply line 18 by 30% (.30)	19	
20	Kilowatt capacity of property on line 18 above ▶ ______ x $1,000	20	
21	Enter the smaller of line 19 or line 20		21
22	Credit carryforward from 2008. Enter the amount, if any, from your 2008 Form 5695, line 28		22
23	Add lines 17, 21, and 22		23
24	Enter the amount from Form 1040, line 46, or Form 1040NR, line 43	24	
25	**1040 filers:** Enter the total, if any, of your credits from Form 1040, lines 47 through 50; line 11 of this form; line 12 of the Line 11 worksheet in Pub. 972 (see instructions); Form 8396, line 11; Form 8839, line 18; Form 8859, line 11; Form 8834, line 22; Form 8910, line 21; Form 8936, line 14; and Schedule R, line 24. **1040NR filers:** Enter the amount, if any, from Form 1040NR, lines 44 through 46; line 11 of this form; line 12 of the Line 11 worksheet in Pub. 972 (see instructions); Form 8396, line 11; Form 8839, line 18; Form 8859, line 11; Form 8834, line 22; Form 8910, line 21; and Form 8936, line 14.	25	
26	Subtract line 25 from line 24. If zero or less, enter -0- here and on line 27		26
27	**Residential energy efficient property credit.** Enter the smaller of line 23 or line 26		27
28	Credit carryforward to 2010. If line 27 is less than line 23, subtract line 27 from line 23	28	

Part III **Current Year Residential Energy Credits**

29	Add lines 11 and 27. Enter the result here and on Form 1040, line 52, or Form 1040NR, line 48, and check box c on that line	29

Form **5695** (2009)

FIGURE 8-3 *Continued*

include labor costs for onsite preparation, assembly or original system installation, and for piping or wiring to interconnect a system to the home. If the federal tax credit exceeds tax liability, the excess amount may be carried forward to the succeeding taxable year. The excess credit can be carried forward until 2016, but it is unclear whether the unused tax credit can be carried forward after then. The maximum allowable credit, equipment requirements and other details vary by technology, as outlined below.

- There is no maximum credit for systems placed in service after 2008. The maximum credit is $2000 for systems placed in service in 2008.
- Systems must be placed in service on or after January 1, 2008, and on or before December 31, 2016.
- The geothermal heat pump must meet federal Energy Star program requirements in effect at the time the installation is completed.
- The home served by the system does *not* have to be the taxpayer's principal residence.

Commercial Tax Credits and Incentives

Federal tax credits for commercial property are seemingly rather anemic by comparison. But appearances are not what they seem. When we first started the Web site for geothermal air-conditioning, I saw the blanket statement on DSIRE:

> The federal business energy investment tax credit available under 26 USC § 48 was expanded significantly by the Energy Improvement and Extension Act of 2008 (H.R. 1424), enacted in October 2008. This law extended the duration—by eight years—of the existing credits for solar energy, fuel cells and microturbines; increased the credit amount for fuel cells; established new credits for small wind-energy systems, geothermal heat pumps, and combined heat and power (CHP) systems; extended eligibility for the credits to utilities; and allowed taxpayers to take the credit against the alternative minimum tax (AMT), subject to certain limitations. The credit was further expanded by The American Recovery and Reinvestment Act of 2009, enacted in February 2009.

For geothermal heat pumps the credit is equal to 10% of expenditures, with no maximum credit limit stated. The system must have been placed in service after October 3, 2008. Interestingly, this credit also covers geothermal systems for electricity generation, with equipment qualifying up to, but not including, the electric transmission stage.

Between all of the legal verbiage, this looks to the layman like commercial systems get only a paltry 10% credit. When I read these rules back in March 2009, I was not impressed, figuring that we would focus more on residential customers. Then one day I was reading on a forum on Geoexchange.org and had my interest piqued by a comment from an industry professional. He stated something along the lines of, "...it's such a shame that the federal government saved the real tax incentives for the business owners. Once again the rich get richer..."

I was perplexed and determined to find out what I was missing. I saw the Maximum Accelerated Cost Recovery System (MACRS) spreadsheet on the ClimateMaster Web site and downloaded it to decipher the benefits (see Fig. 8-4).

IMPORTANT NOTE: *Businesses can end up netting a* total tax credit of *48%, as a result of a first-year depreciation allowance of 50% of the basis. The remaining basis is then depreciated according to a 5-year MACRS schedule, as the example in Fig. 8-4 demonstrates. (See Chap. 10 for more on calculating paybacks.)*

A geothermal system may also be eligible for the section 179 deduction, in which a small business can immediately write-off 100% of spending in lieu of depreciation, to a maximum of $250,000. Learn more about this in form 6251 from the IRS.

If the tax credit exceeds income tax liability, the loss can be carried back one taxable year, and any remaining balance can be carried forward into future years. Note that a business owner who cannot use the tax credits can explore other options, such as sale leasebacks, partnership flip structures, or energy purchase contracts.

As with residential systems, the credit applies to expenditures for parts and labor. The system must be located within the United States and placed in service between October 4, 2008, and December 31, 2016. Entities that are exempt from income taxes, such as churches and charities, are not eligible.

The Feingold-Ensign Support Renewable Energy Act

U.S. Senators Russ Feingold (D-WI) and John Ensign (R-NV) have proposed a bill that could further incentivize geothermal HVAC by closing what many in the industry feel is a current loophole. The Feingold-Ensign Support Renewable Energy Act would add earth-coupled heat pumps—as well as other smaller-scale technologies—to the "official" list of established renewable energy systems, such as solar and wind power. That's important because Congress is already considering a new federal requirement, called the Renewable Electricity Standard (RES), which would mandate that utilities

Business Credit Examples

New Construction Example

A corporation spends $1,000,000 to install a geothermal heat pump system in its new office building. They moved into the building during the 4th quarter of 2009. The corporation is in a 40% tax bracket when state income tax is included.

2009 Tax Credit:	$1,000,000 x 10%	= $100,000
Depreciable Basis:	$1,000,000 – ($100,000/2)	= $950,000
2009 Bonus Tax Benefit:	$950,000 x 50% bonus x 40% tax rate	= $190,000
2009 MACRS Tax Benefit:	$475,000 x 5% Q4 MACRS x 40% tax rate	= $9,500
2010 " " "	$475,000 x 38% MACRS x 40% tax rate	= $72,200
2011 " " "	$475,000 x 22.80% MACRS x 40% tax rate	= $43,320
2012 " " "	$475,000 x 13.68% MACRS x 40% tax rate	= $25,992
2013 " " "	$475,000 x 10.94% MACRS x 40% tax rate	= $20,786
2014 " " "	$475,000 x 9.58% MACRS x 40% tax rate	= $18,202

Total Tax Saving over 5 Years: $480,000

Retrofit Example

A corporation has an existing building that uses a water-loop heat pump system with a boiler and cooling tower. They spend $500,000 to remove the boilers, install a geothermal heat exchange loop, and upgrade their heat pumps to high-efficiency geothermal models. They started the project in 2009 and it became operational in the 1st quarter of 2010. The corporation is in a 40% tax bracket when state income tax is included.

2010 Tax Credit:	$500,000 x 10%	= $50,000
Depreciable Basis:	$500,000 – ($50,000/2)	= $475,000
2010 Bonus Tax Benefit:	Bonus depreciation expires in 2009 (unless it is extended)	
2010 MACRS Tax Benefit:	$475,000 x 35% Q1 MACRS x 40% tax rate	= $66,500
2011 " " "	$475,000 x 26% MACRS x 40% tax rate	= $49,400
2012 " " "	$475,000 x 15.60% MACRS x 40% tax rate	= $29,640
2013 " " "	$475,000 x 11.01% MACRS x 40% tax rate	= $20,918
2014 " " "	$475,000 x 11.01% MACRS x 40% tax rate	= $20,919
2015 " " "	$475,000 x 1.38% MACRS x 40% tax rate	= $2,622

Total Tax Saving over 5 Years: $240,000

Replacement Units Example

A corporation spends $100,000 to install new geothermal heat pumps in its existing building. The geothermal heat pumps are replacing older geothermal heat pumps that were originally installed in 1992. The project is completed in the 3rd quarter of 2009. The corporation is in a 40% tax bracket when state income tax is included.

2009 Tax Credit:	$100,000 x 10%	= $10,000
Depreciable Basis:	$100,000 – ($10,000/2)	= $95,000
2009 Bonus Tax Benefit:	$95,000 x 50% bonus x 40% tax rate	= $19,000
2009 MACRS Tax Benefit:	$47,500 x 15% Q3 MACRS x 40% tax rate	= $2,850
2010 " " "	$47,500 x 34% MACRS x 40% tax rate	= $6,460
2011 " " "	$47,500 x 20.40% MACRS x 40% tax rate	= $3,876
2012 " " "	$47,500 x 12.24% MACRS x 40% tax rate	= $2,326
2013 " " "	$47,500 x 11.30% MACRS x 40% tax rate	= $2,147
2014 " " "	$47,500 x 7.06% MACRS x 40% tax rate	= $1,341

Total Tax Saving over 5 Years: $48,000

FIGURE 8-4 The U.S. federal tax incentives for commercial applications can have a total value of up to 48% of the system cost. In many cases, this brings the adjusted system cost to that of a conventional alternative. (*ClimateMaster*)

obtain a fixed portion of their power from renewable sources. This kind of law has already been on the books in California, Colorado, and other states.

Supporters believe the Feingold-Ensign plan would spur utilities to come on board with what is called *thermal distributions*, which allow the federal government to back loans from the utilities for geothermal systems for homes and businesses. This is better for all of us because the funds become a loan rather than a credit, and the consumer still is able to take advantage of a net-positive cash flow situation. The utility would bill the customer for the loan on their utility billing, at a rate of repayment favorable when compared to the old utility payment with the older, inefficient air-conditioning system.

Home Star (Cash for Caulkers)

Building on the smashing success of the Cash for Clunkers program, in which 680,000 old cars were exchanged for more efficient new ones, the U.S. federal government has proposed the Home Star Energy Retrofit Act, aka *Cash for Caulkers*. This would establish a rebate program for projects that improve energy efficiency in homes, such as insulation and new appliances. Supporters say it would create 170,000 jobs and reduce home energy costs by $10 billion over 10 years.

The bill would authorize $5.7 billion over two years, and an estimated 3 million households are expected to participate. These rebates would be in addition to current tax credits. As currently conceived, the program would include the following levels:

- *Silver Star*, providing up to $4000 in instant rebates. Currently, geothermal air conditioners are left out, although advocates are working hard to change this. It does, however, fund many envelope upgrades that complement geothermal air conditioner retrofits.

- *Gold Star* would provide up to $8000 in instant rebates. Geothermal air conditioners are eligible.

Rural Energy Savings Program Act

The proposed Rural Energy Savings Program Act could become another potential incentive for geothermal systems or other clean technologies. The program would provide electric cooperatives access to $4.9 billion in zero-interest loans through the USDA's Rural Utilities Service. Cooperatives would then make this money available to consumer members in the form of microloans, with an interest rate of no more than 3%, and which can be paid back primarily through savings on their electric bills.

Essentially, a homeowner would present a proposal for a geothermal HVAC system to the utility, and if it were accepted, the utility would fund the job. The consumer would be charged a monthly amount on the utility bill. However, because the homeowner is saving on electrical consumption, the amount charged over a 10-year period to cover the loan should be less than what was previously paid on utility bills. So why not participate?

The program:

- Would be administered through electric co-ops (these are less and less common, in favor of privatized corporate utilities).
- Would offer 10-year low-interest (at a maximum of 3%) financing for energy efficiency retrofits.
- Would be paid for by consumers on electric bills that should end up being less than what was being paid previously.
- Has strong bipartisan and bicameral support.
- Makes geothermal air-conditioning eligible.

State and Local Incentives

Many state and local governments offer incentive programs for geothermal HVAC, including sales and property tax exemptions, income tax credits, and grants. A great place to check availability in the United States is DSIRE, the Database of State Incentives for Renewables and Efficiency Web site (www.dsireusa.org). In the rest of the world, a good place to get started is on RenewableEnergyWorld.com.

At least 28 states have regulatory incentives for geothermal air conditioners that encourage adoption of the technology in public buildings. As stated previously, up to 73% of the points needed to attain a LEED certification can be earned through a geothermal system and related energy controls. This is significant because some states are pushing for certified green buildings for government facilities.

PACE Funding

How many times have you been in the following situation: A salesperson or a Web site has just piqued your interest in a seemingly remarkable product that offers to change your life in a good way. Everything about it is win-win: It increases your comfort, productivity, and ability to enjoy life. . . . It saves you money, helps everyone around you in some way, and it's the right thing to do. But the initial cost is just too much. The bank won't finance you because of that second mortgage, you don't have enough money left on those high-interest credit cards, and even though it's going to provide a net-positive cash flow situation, you can't get the finance company to look past your just-below-perfect credit score.

What if I changed the dream sequence to: The salesperson tells you your credit score doesn't matter. You'll be financed for 20 years at a prime interest rate. The loan stays with whomever owns the property. There is no money down. You still get all of the federal and local incentives. Does this sound too good to be true? It's becoming a reality, and it's called Property Assessed Clean Energy (PACE). PACE (see http://pacelegislation.com for more info) is a relatively recent form of land-secured financing aimed at promoting energy efficiency and renewable energy. Historically, land-secured plans were designed to pay for improvements in the public interest, such as drainage systems, street lighting, and sidewalks. Supporters of PACE point out that the incentives help our communities reduce their environmental impact, lower stress on the electrical grid, and stimulate green jobs.

As of this writing, 21 states have PACE legislation signed into law (several of these have come on board in 2010). In order to administer the program, a regional "district" must be set up. Notably, while many land-secured programs have historically been mandatory in specified districts, PACE is voluntary property by property (and it stays with the property, not the individual). The United States already has more than 37,000 land-secured districts, and increasing numbers of PACE districts are being established, including more than 100 in Florida and appreciable numbers in California, New York, Colorado, and elsewhere.

The funding implement of PACE is really a bond, in which the proceeds are lent to commercial and residential property owners to finance energy projects, such as geothermal systems. Recipients repay these loans over the assigned term (often 15 or 20 years) via an annual special assessment lien on their property, which is collected on their property tax bill. PACE bonds can be issued by municipal financing districts or finance companies (e.g., banks).

PACE programs reduce risks and provide distinct value for mortgage lenders. The *Appraisal Journal* recently pointed out that a $1000 decrease in a home's energy consumption can increase the value of the home significantly. Homes that have been rated as green or LEED-certified sell for up to 10% more than similar homes with standard efficiency measures. Finally, lowered energy bills decrease the risk of default by a property owner.

In October 2009, the White House issued underwriting criteria for PACE funding in order to prevent assistance being given to work that doesn't end up producing real energy efficiency results. For example, a solar array installed in a yard with too few hours of sunlight may be disqualified as could a geothermal air conditioner with the wrong type of ground loop for the soil.

PACE can provide a powerful way for a property owner to afford a green energy system, without needing to have all the money upfront. By enjoying lower utility bills, a participant will have more money on hand to pay down the financing of the project once it's time to settle up on the annual property tax bill.

Summary

In this chapter, we have learned about many of the tax credits and incentives that are currently available for geothermal HVAC systems, which help reduce the payback period and initial costs. The U.S. federal government offers generous breaks for residential and commercial systems. Many international, state, and local agencies also offer benefits to encourage this clean, green technology.

Perhaps not surprisingly, the legislative landscape is constantly changing when it comes to earth-coupled technologies, particularly as awareness of its many benefits continues to build. We expect to see a number of new incentives come online in the near future in various regions of the world, especially as more leaders are interested in stimulating so-called *green jobs*, which drive economic progress as well as sustainable technologies.

CHAPTER 9

Understanding Geothermal Project Proposals

We all want progress, but if you're on the wrong road, progress means doing an about-turn and walking back to the right road; in that case, the man who turns back soonest is the most progressive.

—C.S. Lewis

When I started doing geothermal HVAC in 1990, I used a version of CAD (computer-aided design) to do layout of projects in the planning stage. For example, if I was putting in a slinky horizontal ground loop system, I needed to illustrate clearly what portion of the yard I needed to use.

I have a firm policy that we are to leave a yard not just in ample shape, but better than we found it. As part of a standard package, anything that is damaged is budgeted for and replaced, right down to new sod, mulch in the flowerbed, and some little purple pansies touched with yellow-gold. To make matters even better, we've found great success with local horizontal boring companies, who offer great pricing to run piping all over without any trenching. This has been a marvelous blessing, as was confirmed by several recent customers, who told us that they could not believe how painless the whole process was (Fig. 9-1). Contrast that with those who sometimes complain that geothermal installers destroyed their entire yard.

Egg Geothermal has also formed a tight alliance with a reputable large-scale driller. We have several backup companies, but we rarely have to use them, and we've had great success with staying on budget and on time. The magic combination for us has been to work with separate companies for management and drilling. And, of course, to do ample planning before the first shovel is put in the ground.

That process begins, and is guided by, the proposal. Let's take a closer look.

FIGURE 9-1 A geothermal HVAC system is a big investment. Make certain the proposal is clear and concise. Do not assume that a contractor will do something that isn't spelled out on paper. (*Photo by Egg Systems*)

Typical Geothermal HVAC Proposals

A typical job proposal may be written on a standard one-page sheet that has a few fill-in-the-blank areas and an estimated final amount, with little or no breakdown of costs. You'll notice that a proposal of this nature seems formal and binding enough, yet it has no recourse statements. The lawyer that incorporated Egg Systems in 1989 told me that the "what if" clauses are important because they give the teeth needed to make a contract enforceable.

Poor Proposals

There are those who say that their word is their bond, and that a handshake means everything. Though these individuals may have the very best intentions to do a noble job, interpretation is perhaps 50% of the finished product and rationalization is the other 50%. I can tell you that the best of intentions from the contractor can be reduced to rationalizing down to the builder-grade model after he has a look at the cost overruns that are accruing, and gets anxious to move on to other projects that he may be waiting to start because they are more lucrative.

It happens like this: There are normally three or more grades of equipment eligible for tax credits from a manufacturer. After a positively fabulous meeting for the third straight time, an air of good feeling gives way to a simple handshake, and a standard job invoice or proposal that says something like "four-ton, top-of-the-line geothermal air-conditioning system with all accessories for the Richards' home, $32,000." You ask what about that back bedroom that gets too warm? And he or she says that the new unit has a variable speed blower that provides increased air pressure, which will take care of the problem. You ask about how long you'll be without air, and they say two days, tops. You don't want to insult your new buddy, so you find yourself wanting to believe that it will just magically turn out A-OK.

At this point it's a good idea to pull out a quarter and call it in the air: Heads I win, tails you lose. Actually, many contractors are fairly dependable, and even old-school honest. But many more have all the noble intent in the world, and little ability to back it up. I recently had an addition done to my home, and after many years of dealing with contracts, I still missed some items that ended up costing me extra money because they weren't properly detailed, such as the bathroom fixtures. The proposal specified "custom* fixtures selected by owner and supplied by electrician." When the time came, we brought in the part number and picture of what we wanted, but it was not on the custom list that the electricians had provided. The asterisk referred to a Web site that showed a builder-grade selection of fixtures straight out of the $7- to $12-section of Home Depot.

Many outfits can really get you on the trim and the details. Take the time to spell out everything up front. Maybe even take an evening and watch the movie *Money Pit* from the 1980s. I still chuckle to myself when I think about the general contractor who drove up in the Jaguar, stepped out and said, "Quick, write me a check for $5000 before I change my mind." I know that it can happen to the best of contractors.

Back to our geothermal scenario, your contractor finds funds a little tight due to some circumstances unrelated to your job, and she downgrades you from a 30 EER unit to a 22 EER unit, rationalizing that she gave you too good a deal for the money to begin with, and that you can upgrade easily later if you really want to. She rationalizes that a 22 EER unit has an equivalent SEER rating of 27, outpacing the best air-source equipment by eight or more points.[1] By the time

[1] More details on EER auditing are in Chapter 11.

she is done with you, you can have a B+ job when you thought you had an A+ job.

Normally, you really don't have the time or the patience to figure it all out. I know I don't. Plus, I really want to believe that I'm being treated fairly, and that everyone is honest. This is especially true if I haven't covered all my bases; subconsciously, I may know I've been shorted, but I don't want to know by how much.

Figure 9-2 shows an actual proposal that we think is missing several key elements. It lacks many specific details and states that a system will be installed "as we outlined verbally," instead of spelling out exact specifications. It has notes scrawled by hand over the form, which lacks professionalism and clarity.

Good Proposals

A good proposal, as in Fig. 9-3, actually involves such detail that the comfort level approaches something on par with the education that you'd get from reading a book like this, and then having the salesperson address each issue in so much depth that you feel as if you are the only person on the face of the Earth to ever have such a system installed. It will involve satellite photos of your home or business with arrows and dashed lines showing the exact routing and location of every penetration, pipe, well, trench, and piece of equipment, along with the ductwork, controls, and valves, all in a tidy CAD drawing, complete with a title block with your name on it. I don't know how everyone feels about this, but when I see a proposal with my name emboldened on it, "Geothermal Design for the Home of Mr. and Mrs. Egg," I immediately feel important to the contractor. At this point, I want to give him or her my business.

This proposal will have pictures of the old equipment and brochures on the exact model and orientation of the new equipment going in. It will have piping layouts throughout the building, details on controls, and fluid GPM requirements. It will spell out any subcontractors involved, such as the electrician, plumber, and driller. He or she will have you, the building owner, sign off on several things, such as the locations of wells, trenches, permission to remove shrubbery, and other items. They'll have you file a notice of commencement and copy you with a permit. Their office will be in constant contact on every detail. You will feel that this is one of the most pleasant experiences of your life . . . if you pick the right contractor.

Do not suppose that if a contractor does not have it together by the first site visit that she will get it together later. It only gets worse. Every time—no ifs, ands, or buts. If you want to know how the contractor in the scenario above was able to present you with such a

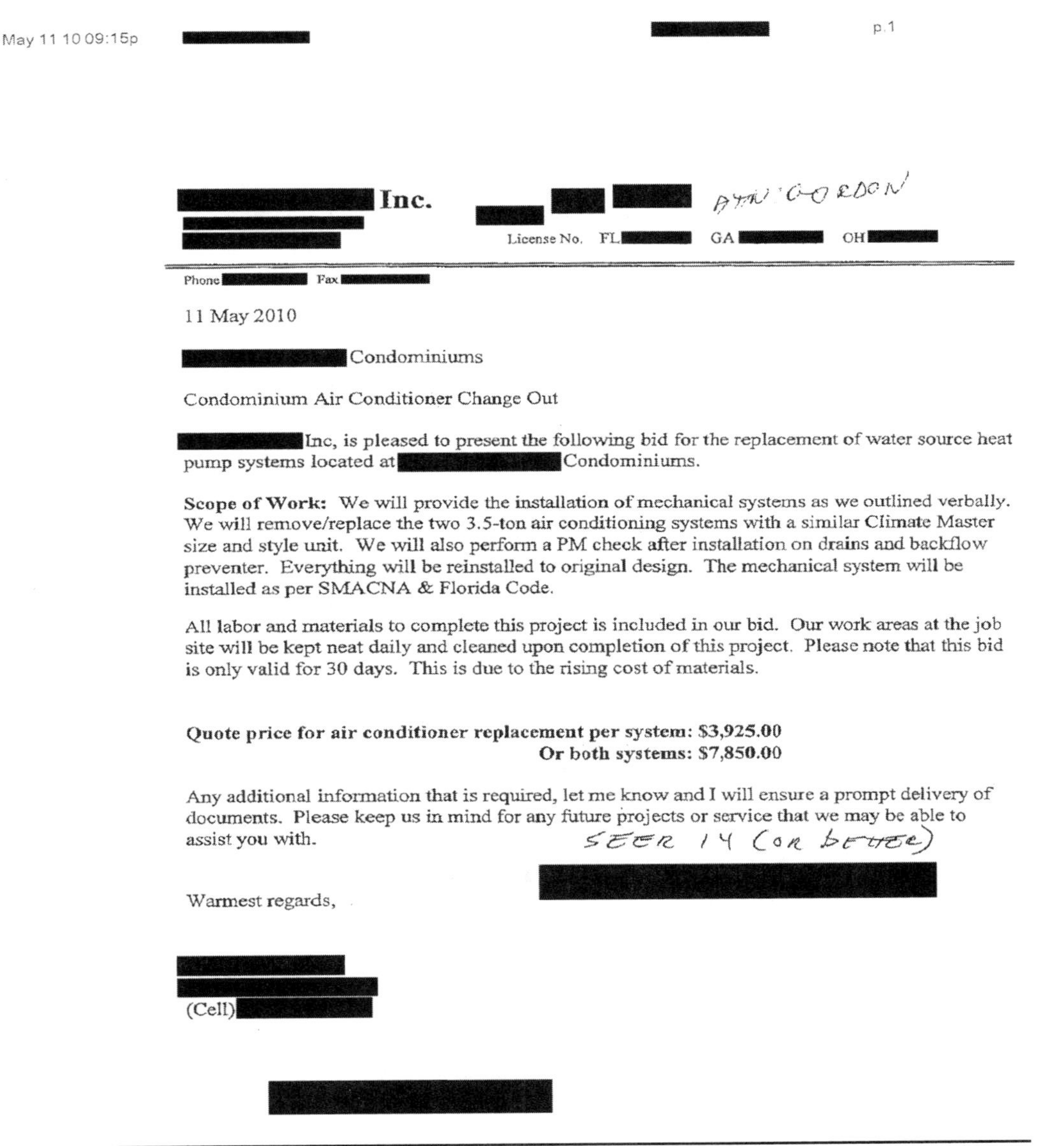

May 11 10 09:15p p.1

Inc. ATTN: GORDON

License No. FL GA OH

Phone Fax

11 May 2010

Condominiums

Condominium Air Conditioner Change Out

Inc, is pleased to present the following bid for the replacement of water source heat pump systems located at Condominiums.

Scope of Work: We will provide the installation of mechanical systems as we outlined verbally. We will remove/replace the two 3.5-ton air conditioning systems with a similar Climate Master size and style unit. We will also perform a PM check after installation on drains and backflow preventer. Everything will be reinstalled to original design. The mechanical system will be installed as per SMACNA & Florida Code.

All labor and materials to complete this project is included in our bid. Our work areas at the job site will be kept neat daily and cleaned upon completion of this project. Please note that this bid is only valid for 30 days. This is due to the rising cost of materials.

Quote price for air conditioner replacement per system: $3,925.00
Or both systems: $7,850.00

Any additional information that is required, let me know and I will ensure a prompt delivery of documents. Please keep us in mind for any future projects or service that we may be able to assist you with.

SEER 14 (OR BETTER)

Warmest regards,

(Cell)

FIGURE 9-2 In this proposal, the only certain conclusion the customer can make is that he or she will get some type of geothermal air conditioner rated at 14 SEER or better . . . and that it will be installed as outlined verbally. We don't think that is enough detail.

Geothermal Proposal

Date: 5/7/2010

Customer Information		**Job Information**	
Name:	[redacted]	Name:	[redacted]
Address:	[redacted]	Address:	Same
City,ST Zip:	Dunedin, FL 34698	City,ST Zip:	-
Phone:	[redacted]	Phone:	-
Email:	[redacted]	Email:	-

Thank you for choosing Egg Geothermal for your geothermal air conditioning needs. We appreciate the opportunity to be of service and are confident that you will be completely satisfied with your geothermal investment. This proposal is comprised of 4 Sections including Geothermal Heating & Air Conditioning Equipment, Geothermal Pool Heating & Cooling Equipment, the required Water Loop System, and Other Required Items or Modifications that are part of this agreement. Non Applicable sections will be noted.

Section I: Geothermal Heating and Air Conditioning Equipment

Egg Geothermal hereby proposes to install a new high efficiency geothermal air conditioning system at the address listed above. This system meets the qualifications for the Federal Tax Credit that is available for geothermal air conditioning/heat pump systems. The following equipment, ductwork, venting, materials, taxes and labor are included in the proposal:

Description	Quantity
3 ton ClimateMaster Tranquility 27TT038	1

The Tranquility 27 TT Series is Variable Speed with a Two Stage Compressor which make it one of the most efficient systems available to provide for your comfort needs. The system comes complete with Cupro-Nickel, E-Coated coils and is equipped with the HWG - Hot Water Generating system. The system is also environmentally friendly with R410A Earthpure Refrigerant.

Standard Installation Includes:

Duct Modification - Minor duct modifications included to match existing supply and return ductwork to the new system. All new connections will be taped and mastic sealed to prevent air loss.
Thermostat - New Thermostat provided with Humidistat Control
Motorized Valves - Motorized Valve piped on exiting water line to keep system under pressure. Slow closing mechanism prevents water hammer.
Flow Raters - Flow Raters installed to regulate total gpm of water flowing through system.
PT Ports - Pressure & Temperature ports installed on incoming and outgoing water lines.
Ball Valves - Ball valves included on both entering and exiting water lines, include clean out ports.
Plumbing - All Water Lines run with CPVC and insulated as required by code.
Electrical - High Voltage wiring run to pump. New disconnect and whip provided at unit.
Equipment Removal - Removal of the existing equipment from the premises.
Landscaping - Every effort will be made to protect landscaping. There is a lot of ground work required in this project and landscaping will be maintained and restored at conclusion of work.

7906 Leo Kidd Avenue Port Richey, FL 34668 727.848.7545 Office, 727.848.0134 Fax
Lic. #: CAC053846, CMC1249626 www.egggeothermal.com 1 of 3

FIGURE 9-3 A more detailed proposal will include specific guidelines, schedules, and payback scenarios. Cost comparisons and exact equipment specifications, along with installation locations, will be spelled out. (*Egg Systems*)

Permits - All work performed in accordance with code requirements, all necessary permits obtained, notice of commencement filed and inspection completed.

Green Egg Club- 2 years of our Green Egg Club Membership are included with every installation. Includes priority service and annual preventative maintenance visit.

Warranty - The Manufacturers Warranty is 10 Year Parts and Labor. The installation is warrantied for 1 year.

Special Job Notes:

All Work Performed in a neat professional manor by certified and trained technicians. Sweeping, dusting, and vacuuming will be accomplished at the conclusion of each days work and all debris removed from the premises. Our technicians will treat your home with respect. They will not smoke or swear on your property and will maintain a neat, professional appearance.

Geothermal Equipment Investment	$	18,869
Egg Geothermal Discount:	$	1,887
Total Geothermal Equipment Investment	**$**	**16,982**

Section II: Geothermal Pool Heating and Cooling Equipment **N/A**

Section III: Water Loop System

Loop Type: Open Loop - Supply Well with VFD Submersible Pump, Injection Well
Well Contractor: Coastal Caisson
Total Wells: 2 **Depth Quoted:** 150 **Casing Depth:** 100

This estimate is based on historical data relative to deep wells that have been installed in this region. We make every attempt to be accurate with the drilling depth quoted. As we do not know exactly what the geographical layout is under your property additional Drilling/Casing may be required. If additional drilling or casing is required the following schedule would apply:

Additional Drilling per Foot: $ 17.00 **Additional Casing per $ 19.00**

Description:
This is not an actual estimate on the required wells. These figures are based on averages in the area for a two well system with Variable Frequency Drive pump. Actual well bid request has been filed and are waiting on final numbers

Total Water Loop Investment: **$ 9,500**

Section IV: Other Required Items or Modifications: **N/A**

Total Project Investment:

Total Geothermal Heating & Air Conditioning Equipment $ 16,982

7906 Leo Kidd Avenue Port Richey, FL 34668 727.848.7545 Office, 727.848.0134 Fax
Lic. #: CAC053846, CMC1249626 www.egggeothermal.com 2 of 3

FIGURE 9-3 *Continued*

Total Geothermal Pool Heating & Cooling Equipment	$ -
Total Water Loop Investment	$ 9,500
Total Other Investment	$ -
Total Investment Amount:	**$ 26,482**

Terms:

60% Down	$ 15,889
40% Upon Completion	$ 10,593

Agreement:

[redacted] Date   Egg Systems Date

7906 Leo Kidd Avenue Port Richey, FL 34668 727.848.7545 Office, 727.848.0134 Fax
Lic. #: CAC053846, CMC1249626 www.egggeothermal.com 3 of 3

Figure 9-3 *Continued*

RENEWABLE • SUSTAINABLE • COMFORTABLE

Return On Investment - Residential Geothermal

Home Owner: [redacted]
Job Info: upgrage to 3 ton geothermal system

Prepared By: J. Fontano
Date: 5/7/2010

Geothermal Investment Information

Total Geothermal Investment: (estimate only, well/loop amount will vary)	$	26,482.10
Estimated Yearly Hot Water Savings:	$	250.00
Federal Tax Credit:	$	7,944.63
Net Geo Investment (after tax credit):	$	18,537.47

Geothermal System Savings

Net Geo Investment (after tax credit):	$	18,537.47
Cost of High Efficiency Standard Eqp: (investment for high efficiency A/C upgrade)	$	12,000.00
Total Additional Investment for Geo Upgrade:	$	6,537.47
10 Year Repair Savings w/ Ext Warranty	$	3,144.47
1st Year Energy Savings	$	1,135.00
Years to Recover Additional Investment:		3.8

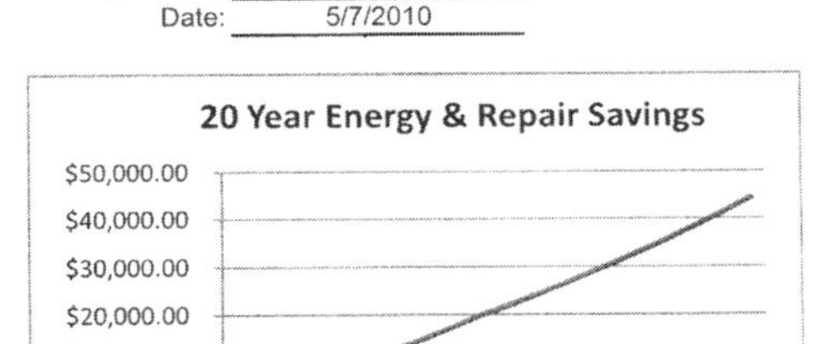

Annual Air Conditioning Energy, Repair, and Free Hot Water Savings

Year	Annual Energy Savings (4% Annual Utility Increase)	Annual Repair Savings (10 Year Warranty)	HWG Hot Water Savings	Total Annual Savings	Cumulative Savings
Year 1	$ 1,135.00	$ 250.00	$ 250.00	$ 1,635.00	$ 1,635.00
Year 2	$ 1,180.40	$ 262.50	$ 260.00	$ 1,702.90	$ 3,337.90
Year 3	$ 1,227.62	$ 275.63	$ 270.40	$ 1,773.64	$ 5,111.54
Year 4	$ 1,276.72	$ 289.41	$ 281.22	$ 1,847.34	$ 6,958.88
Year 5	$ 1,327.79	$ 303.88	$ 292.46	$ 1,924.13	$ 8,883.01
Year 6	$ 1,380.90	$ 319.07	$ 304.16	$ 2,004.13	$ 10,887.15
Year 7	$ 1,436.14	$ 335.02	$ 316.33	$ 2,087.49	$ 12,974.64
Year 8	$ 1,493.58	$ 351.78	$ 328.98	$ 2,174.34	$ 15,148.98
Year 9	$ 1,553.33	$ 369.36	$ 342.14	$ 2,264.83	$ 17,413.81
Year 10	$ 1,615.46	$ 387.83	$ 355.83	$ 2,359.12	$ 19,772.93
Year 11	$ 1,680.08		$ 370.06	$ 2,050.14	$ 21,823.07
Year 12	$ 1,747.28		$ 384.86	$ 2,132.14	$ 23,955.21
Year 13	$ 1,817.17		$ 400.26	$ 2,217.43	$ 26,172.64
Year 14	$ 1,889.86		$ 416.27	$ 2,306.13	$ 28,478.77
Year 15	$ 1,965.45		$ 432.92	$ 2,398.37	$ 30,877.14
Year 16	$ 2,044.07		$ 450.24	$ 2,494.31	$ 33,371.45
Year 17	$ 2,125.83		$ 468.25	$ 2,594.08	$ 35,965.53
Year 18	$ 2,210.87		$ 486.98	$ 2,697.84	$ 38,663.37
Year 19	$ 2,299.30		$ 506.45	$ 2,805.76	$ 41,469.13
Year 20	$ 2,391.27		$ 526.71	$ 2,917.99	$ 44,387.11
Total Savings	**$ 33,798.12**	**$ 3,144.47**	**$ 7,444.52**	**$ 44,387.11**	

This document is for informational purposes only. By providing you with this information Egg Geothermal Systems, Inc. is not providing, nor intending to provide, you or any other reader of this document with legal or tax advice. For legal or tax advice you should consult a lawyer or CPA. Neither Egg Geothermal Systems, Inc. nor its affiliates or consultants shall be responsible for your use of this document or for any damages resulting therefrom.

FIGURE 9-3 *Continued*

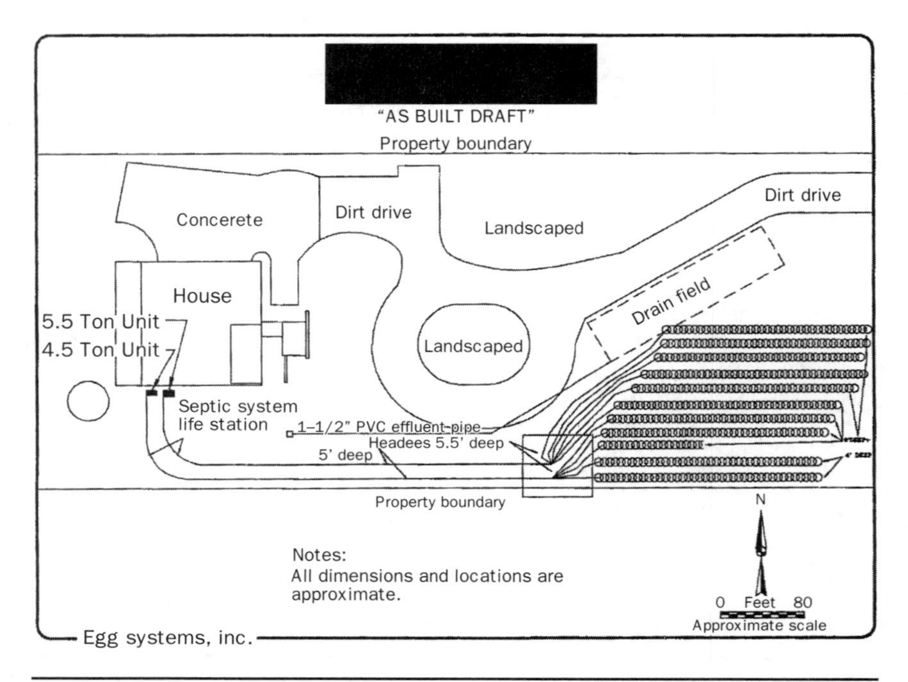

FIGURE 9-3 *Continued*

customized proposal, we'll tell you. After you inquired on the Web, or answered one of the magazine ads in *Popular Science*, a friendly assistant may have called you to get some information. He or she would have made you feel good about having made the inquiry. You would have given your address, the square footage of your home, whether or not you have a pool, what fuel source you use, some information on the age, and perhaps the model number of your existing system, and you may have had a chance to share what your goals for energy efficiency and savings are. This person would ask if you're having any problems with your current air conditioner, so he or she could properly prioritize your needs. If your system were inoperable for any reason, you would likely be informed that they would expedite a service technician to you.

If not, you might be informed that due to the incredible response to geothermal air-conditioning requests, they are about four weeks behind on appointments, but that should you have any problems with your current system during this process, you would be taken care of. I feel pretty safe in saying that I would be optimistically cautious of a company that is able to be out in the next day

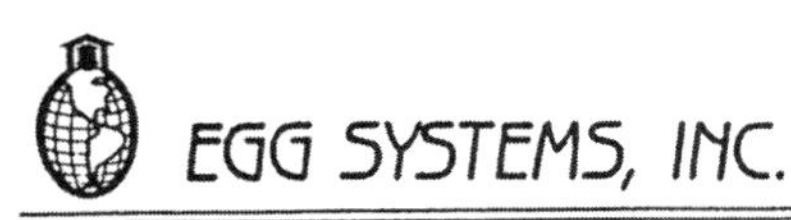

"Mechanical Systems for an Energy Conscious World"

GEOTHERMAL AIR CONDITIONING SYSTEM INSTALLATION PROPOSAL

Date: ________________

Thank you for the opportunity to propose the geothermal air conditioning system. This proposal includes:

» Installation of a ____ ton Addison water source heat pump model WPG____;
» Construction and installation of __________ loop (geothermal heat exchanger) placed ____;
» Installation of piping from loop to water source heat pump, including pump station and valve box;
» Installation of heat recovery piping and pump for hot water recovery;
» Installation of a standard heat pump thermostat; and
» Installation of all ductwork including ___ supply registers and ___ return registers.

The total cost to perform the above-listed work is $ _________ plus tax, which can be broken down as follows:

Equipment	
» Water Source Heat Pump	$_______
» Loop Supplies	$_______
» Duct and Installation Supplies	$_______
Subtotal	$_______
Subcontractor	
» __________________	$_______
Labor	
» Installation Labor	$_______
TOTAL COST	**$_______**

If you find this proposal acceptable, please sign and date below. Thank you for you consideration.

Sincerely,

Jay Egg

Accepted by: ____________________________

Date: ____________________________

Scheduled Start Date: ____________________________

Our New Address Is:
EGG SYSTEMS, INC.
250 Burbank Rd., Suite 2
Oldsmar, Florida 34677

35246 U.S. 19 N., Suite 299 • Palm Harbor, FL 34684 • (813) 789-3841

FIGURE 9-4 In the early 1990s, Egg Geothermal performed so many geothermal proposals that we used the "fill-in-the-blank" proposal shown. This was not a good idea, as we learned with experience... geothermal air-conditioning proposals need more detail and preengineering. (*Egg Systems*)

or two. That might mean that they are just getting into the technology. Normally, the best contractors are the busiest contractors for a reason.

Moving along in this proposal process, you would receive a call by a sales engineer, who would get a few more details and talk about some qualifications and some financial questions. Then you would be invited to a live, online sales presentation, customized to your needs and your requests. By the time you finish viewing this presentation and talking with the sales engineer (these folks have to really know their business), you will be a wealth of knowledge and have abiding confidence in the concept of geothermal HVAC and peripheral technologies. You will have a date for a site visit, and you will know about what the system will cost. Hopefully, you will be about ready to buy.

Of course, the sales rep who comes out to do the presentation on-site will be finding all of the obstacles those satellite images missed. That might be a too-small closet, or a necessary relocation of the air handling assembly . . . maybe a duct or filtration upgrade and some ultraviolet light to keep the bacteria growth down.

Never fear, all these additional issues are covered (as of the date of this printing) in the federal stimulus, as long as they are part of an Energy Star-rated geothermal air-conditioning system that qualifies under the 2009 package. So, when you were surprised by that great detailed proposal with your name on it, the contractor had already done her homework, and you are seeing the fruits of hours of labor and engineering. Please be careful and help us to help you. Always get a detailed, engineered plan by a reputable company.

Summary

Figure 9-4 shows the proposal form I used when I first got started in this business in the early 1990s. As you can see, I have learned a lot over the past decades, and I can now offer customers more sophisticated estimates, with more detailed descriptions of expected costs and payback periods. A good job proposal should leave nothing up to chance or misinterpretation and should spell out the exact equipment to be used. You are spending a lot of money on a new system, whether for a residential or commercial application, and you deserve a detailed roadmap of the process.

CHAPTER 10

How to Calculate Your Payback

At 15 years old, I got my first job. I was supposed to be 16 to be allowed to work as a dishwasher at the Harvey House Restaurant in Barstow, California.[1] Apparently, they needed me, and my birthday was in September anyway, and it was June. I logged a lot of hours that summer. . . . I worked my way up to a line cook, which they said could not be done in so short a time. I was making well over minimum wage.

A little later that year, Bishop Stone secured a job for me with an electrical contracting company that was working on our church remodel, and thus started my career in the trades. By the age of 18, I was a fully qualified commercial journeyman electrician, and was working on solar power test sites in the Mojave Desert.

I filed my own taxes those first few years on a 1040EZ form. It took me all of ten minutes, including the walk to the mailbox. But fast-forward 30 years, and I am involved in multiple businesses and investments that require teams of accountants. I don't even pay my own bills because I would get into trouble. You know the type: I could never keep track of what was automatically withdrawn or who needs to be paid online or by check. It takes a monthly audit just to figure out how much I earned so that I can pay tithing to the church.

So now I understand the importance of certified public accountants. Not only do they complete the necessary tax forms, but they also help with business start-ups and give advice all along the way. My team at Egg Geothermal has held meetings with several local certified public accountant firms to help them understand the new geothermal tax credits and incentives that are available, so they can better advise their customers.

[1] By the way, about my first job at the Harvey House, you can't imagine my surprise when I returned 25 years later (Fig. 10-1) with my wife and children to show them my first place of employment, and Barbara and Vera were still there. Very cool!

Figure 10-1 Twenty-five years after gaining valuable business experience at Harvey House Restaurant, Jay Egg visits with the staff he remembers in Barstow, California. (*Egg Systems*)

Figure 10-2 A ClimateMaster technician installs a vertical space-saving heat pump unit in a New York City apartment building. (*Photo by ClimateMaster*)

The purpose of this chapter is to illustrate that the purchase of a geothermal system will provide a net positive cash flow scenario (Fig. 10-2). The staff at Egg Geothermal has worked hard to try to simplify the process of making the relevant accounting calculations. We've listened to feedback from hundreds of customers, and we hope this is helpful.

Determining ROI on Residential Systems

The spreadsheet in Fig. 10-3 shows return on investment (ROI) calculations from the proposal prepared for a real Egg Geothermal customer. As you will see, we are able to summarize the cost of the geothermal system, the associated costs of electricity for standard or existing systems, the costs and savings for domestic hot water, and arrive at a timeframe to recoup the investments.

Using the starting point of the cost of a geothermal system to a customer—in this case $42,438—we are able to subtract a conservative estimate for domestic hot water savings ($500 a year in this example). The next line indicates the federal tax credit, $12,731 for this customer. The difference is then indicated as the net geothermal investment, which is $29,707 here.

Since many customers are calling because their current HVAC system is in need of repair, or because they are building a new home, we like to add a comparison figure of the amount it would cost for a high-efficiency standard air-conditioning system. This is as if to say, "You were in the market to purchase an efficient system, so let's look at what you would have spent anyway if you didn't choose geothermal." In our example, that would be $22,000.

The 10-year repair savings listed ($6289) is an indicator of the average dollars we have seen customers spend to keep their conventional air conditioners maintained and running well for that time period. This often varies by unit size, brand, and type. These figures include annual maintenance and coil cleaning, as well as repair and other incidental costs.

The annual energy savings are calculated through proprietary programs that are offered by various manufacturers and software providers. The most accurate of these programs take many factors into account, including the square footage of the home, the number of people living there, the construction age and quality, the regional climate, and past energy consumption habits and trends. We have found many of these to be extremely accurate, but we prefer to err on the conservative side so as to allow a margin for a pleasant surprise factor. For this customer, we figured on $1617 for the first year, although as Fig. 10-3 shows, this increases up to $3407 by the 20th year.

The difference of these values is taken, and a net payback is factored in on the bottom line, indicated as, "years to recover additional investment." In this real-world example, that is only 2.9 years. The

RENEWABLE • SUSTAINABLE • COMFORTABLE

Return On Investment - Residential Geothermal

Home Owner: [redacted]
Job Info: Geothermal Upgrade - (2) 3 ton systems

Prepared By: J. Fontano
Date: 5/3/2010

Geothermal Investment Information

Total Geothermal Investment: (estimate only, well/loop amount will vary)	$	42,438.00
Estimated Yearly Hot Water Savings:	$	500.00
Federal Tax Credit:	$	12,731.40
Net Geo Investment (after tax credit):	$	29,706.60

Geothermal System Savings

Net Geo Investment (after tax credit):	$	29,706.60
Cost of High Efficiency Standard Eqp: (investment for high efficiency A/C upgrade)	$	22,000.00
Total Additional Investment for Geo Upgrade:	$	7,706.60
10 Year Repair Savings w/ Ext Warranty	$	6,288.95
1st Year Energy Savings	$	1,617.00
Years to Recover Additional Investment:		2.9

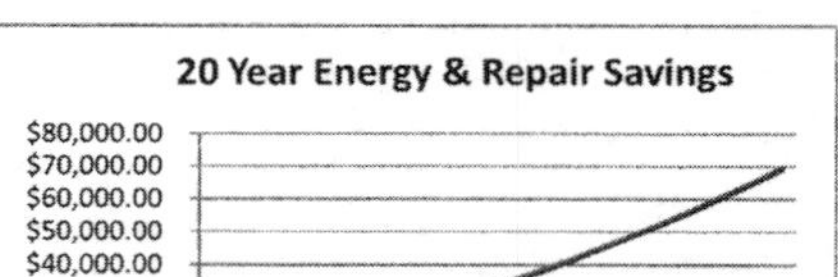
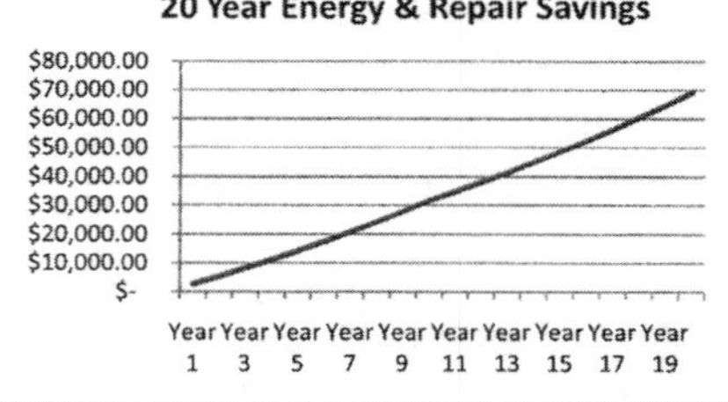

Annual Air Conditioning Energy, Repair, and Free Hot Water Savings

Year	Annual Energy Savings (4% Annual Utility Increase)	Annual Repair Savings (10 Year Warranty)	HWG Hot Water Savings	Total Annual Savings	Cumulative Savings
Year 1	$ 1,617.00	$ 500.00	$ 500.00	$ 2,617.00	$ 2,617.00
Year 2	$ 1,681.68	$ 525.00	$ 520.00	$ 2,726.68	$ 5,343.68
Year 3	$ 1,748.95	$ 551.25	$ 540.80	$ 2,841.00	$ 8,184.68
Year 4	$ 1,818.91	$ 578.81	$ 562.43	$ 2,960.15	$ 11,144.83
Year 5	$ 1,891.66	$ 607.75	$ 584.93	$ 3,084.34	$ 14,229.17
Year 6	$ 1,967.33	$ 638.14	$ 608.33	$ 3,213.79	$ 17,442.97
Year 7	$ 2,046.02	$ 670.05	$ 632.66	$ 3,348.73	$ 20,791.69
Year 8	$ 2,127.86	$ 703.55	$ 657.97	$ 3,489.38	$ 24,281.07
Year 9	$ 2,212.98	$ 738.73	$ 684.28	$ 3,635.99	$ 27,917.06
Year 10	$ 2,301.50	$ 775.66	$ 711.66	$ 3,788.82	$ 31,705.88
Year 11	$ 2,393.56		$ 740.12	$ 3,133.68	$ 34,839.55
Year 12	$ 2,489.30		$ 769.73	$ 3,259.02	$ 38,098.58
Year 13	$ 2,588.87		$ 800.52	$ 3,389.39	$ 41,487.96
Year 14	$ 2,692.42		$ 832.54	$ 3,524.96	$ 45,012.92
Year 15	$ 2,800.12		$ 865.84	$ 3,665.96	$ 48,678.88
Year 16	$ 2,912.13		$ 900.47	$ 3,812.60	$ 52,491.48
Year 17	$ 3,028.61		$ 936.49	$ 3,965.10	$ 56,456.58
Year 18	$ 3,149.76		$ 973.95	$ 4,123.71	$ 60,580.29
Year 19	$ 3,275.75		$ 1,012.91	$ 4,288.65	$ 64,868.94
Year 20	$ 3,406.78		$ 1,053.42	$ 4,460.20	$ 69,329.14
Total Savings	**$ 48,151.15**	**$ 6,288.95**	**$ 14,889.04**	**$ 69,329.14**	

This document is for informational purposes only. By providing you with this information Egg Geothermal Systems, Inc. is not providing, nor intending to provide, you or any other reader of this document with legal or tax advice. For legal or tax advice you should consult a lawyer or CPA. Neither Egg Geothermal Systems, Inc. nor its affiliates or consultants shall be responsible for your use of this document or for any damages resulting therefrom.

7906 Leo Kidd Ave., Port Richey, FL 34668 Office: 727.848.7545 Fax: 727.848.0134
Lic #: CAC053846, CMC1249626 www.egggeothermal.com

FIGURE 10-3 A return on investment (ROI) calculation should not only look at energy savings, but also longer lifespan, decreased service hours, elimination of cooling tower equipment and associated treatment, etc. (*Egg Systems*)

graph at the right of the figure illustrates the 20-year savings for repair and energy costs. To see these numbers broken down, observe that by adding up all the energy savings ($48,151), maintenance savings ($6289), and money saved on producing hot water ($14,889), the total amount of money saved by this geothermal system is an impressive $69,329.

The only thing left for the customer to do is look at the real amount they are investing in the system. The exciting thing is that they can set up a payment plan with their financial institution that requires no additional money from their existing monthly budget. This is done by setting the monthly payment amount at less than the anticipated monthly energy savings. In this case, the customer is going to save $176 per month, so the loan might be set for $150 per month. Usually, the resulting payback period is around 10 to 20 years on smaller systems, although it goes much quicker as systems get larger. As with any loan, the customer could pay more to pay it down faster, but the point here is that there is often an option to afford a new system without adding to the household budget. And the customer gets to keep a high-quality system that will far outlive the loan payments. Other than real estate, that just doesn't happen too often!

ROI on Geothermal Pool Heat Pumps

Geothermal pool heat pumps often pay for themselves very quickly. Compared with gas or electric heating, they save 70% to 90% on overall energy costs.

In the example shown in Fig. 10-4, the customer is going to spend an additional $11,541 on the installation of a geothermal pool heat pump. This particular customer spent $10,832 last year for gas to keep the caged pool warm enough to swim all year long. That amount will be reduced to $1411 with the new geothermal pool heat pump, reflecting a cost savings of $9420, or 87%! Note that there are no tax incentives of which we are aware for the purchase or installation of a pool heat pump system. However, it is easy to see that an earth-coupled pool heat pump still has an impressively short payback period—a mere 1.23 years in this case! As a result, increasing numbers of municipal and larger club pools, such as the YMCA, are converting over to this proven technology.

IMPORTANT NOTE: *A pool can rarely be placed on a ground loop heat exchanger due to the extremely high level of thermal extraction. Additionally, a properly engineered condenser water loop can be exponentially advantageous to the performance of the air-conditioning systems in the building, home, or clubhouse that may be adjacent to the pool.*

Prepared By: J. Fontano
Customer Name:
Date: 5/3/2010

Geothermal Pool Heat Pump - Return on Investment

Total Investment	$ 11,541.00
Annual Pool Heating Expense - Other	$ 10,832.15
Est. Annual Geothermal Operating Amount	$ 1,411.76
Estimated Annual Energy Savings	$ 9,420.39
Payback for Geothermal Pool Heat Pump in Years	1.23

This document is intended for informational purposes only using estimated information provided by Aqua Cal (the manufacturer) based on historical data from your geographical area that has been collected over several years. Actual operating amounts and savings vary depending on individual operating habits.

FIGURE 10-4 A commercial or residential pool heater installation often has a payback of about three years or less. (*Egg Systems*)

Calculating Payback Periods for Commercial Geothermal Systems

Commercial systems are basically the same story as residential systems, but they have higher overall incentives, and they often have high penalties for demand or "spikes" in consumption. As a result, geothermal systems for commercial applications almost always have a very impressive payback.

In the example shown in Fig. 10-5, the customer is purchasing some air-conditioning equipment for a new convenience store connected to a fueling station. They are going to pay $76,182 for a geothermal air-conditioning system to cool and heat the space. The federal tax credit is 10%, which on the face sounds like a short change when compared to the residential tax credit of 30%. But as we saw in Chapter 8, the federal government also offers a 5-year accelerated depreciation plan on geothermal equipment as part of the recent stimulus package.

In 2008, before the stimulus effort passed, David Kyle, a member of the Board of Directors of the Air-Conditioning Contractors of America (and owner of Trademasters Service Corp. of Lorton, Virginia), testified before the Treasury Department that, "Currently the tax code treats commercial HVACR [Heating, Ventilation,

Return on Geothermal Investment - Commercial

Home Owner: ███████
Job Info: Geothermal RTU's

Prepared By: J. Fontano
Date: 5/10/2010

Geothermal Investment Information

Total Geothermal Investment: (estimate only, well/loop amount will vary)	$	76,182.42
Estimated Yearly Hot Water Savings:	$	750.00
Federal Tax Credit:	$	7,618.24
5 Year MACRS:	$	28,949.32
Net Geo Investment (after tax credits):	$	39,614.86

Geothermal System Savings

Net Geo Investment (after tax credit):	$	39,614.86
Cost of High Efficiency Standard Eqp: (investment for high efficiency A/C upgrade)	$	39,260.03
Total Additional Investment for Geo Upgrade:	$	354.83
5 Year Repair Savings w/ Ext Warranty	$	2,768.34
1st Year Energy Savings	$	5,122.00
Years to Recover Additional Investment:		0.056

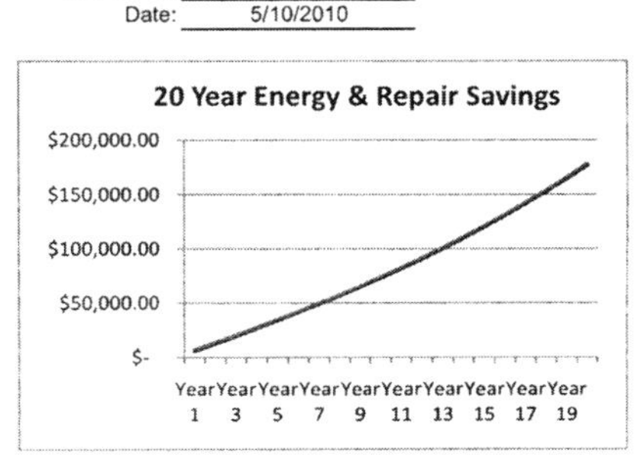

Annual Air Conditioning Energy, Repair, and Free Hot Water Savings

Year	Annual Energy Savings (4% Annual Utility Increase)	Annual Repair Savings (5 Year Warranty)	HWG Hot Water Savings	Total Annual Savings	Cumulative Savings
Year 1	$ 5,122.00	$ 501.00	$ 750.00	$ 6,373.00	$ 6,373.00
Year 2	$ 5,326.88	$ 526.05	$ 780.00	$ 6,632.93	$ 13,005.93
Year 3	$ 5,539.96	$ 552.35	$ 811.20	$ 6,903.51	$ 19,909.44
Year 4	$ 5,761.55	$ 579.97	$ 843.65	$ 7,185.17	$ 27,094.61
Year 5	$ 5,992.02	$ 608.97	$ 877.39	$ 7,478.38	$ 34,572.99
Year 6	$ 6,231.70		$ 912.49	$ 7,144.19	$ 41,717.17
Year 7	$ 6,480.96		$ 948.99	$ 7,429.95	$ 49,147.13
Year 8	$ 6,740.20		$ 986.95	$ 7,727.15	$ 56,874.28
Year 9	$ 7,009.81		$ 1,026.43	$ 8,036.24	$ 64,910.52
Year 10	$ 7,290.20		$ 1,067.48	$ 8,357.69	$ 73,268.20
Year 11	$ 7,581.81		$ 1,110.18	$ 8,691.99	$ 81,960.20
Year 12	$ 7,885.08		$ 1,154.59	$ 9,039.67	$ 90,999.87
Year 13	$ 8,200.49		$ 1,200.77	$ 9,401.26	$ 100,401.13
Year 14	$ 8,528.51		$ 1,248.81	$ 9,777.31	$ 110,178.44
Year 15	$ 8,869.65		$ 1,298.76	$ 10,168.40	$ 120,346.85
Year 16	$ 9,224.43		$ 1,350.71	$ 10,575.14	$ 130,921.99
Year 17	$ 9,593.41		$ 1,404.74	$ 10,998.15	$ 141,920.13
Year 18	$ 9,977.15		$ 1,460.93	$ 11,438.07	$ 153,358.21
Year 19	$ 10,376.23		$ 1,519.36	$ 11,895.59	$ 165,253.80
Year 20	$ 10,791.28		$ 1,580.14	$ 12,371.42	$ 177,625.22
Total Savings	**$ 152,523.32**	**$ 2,768.34**	**$ 22,333.56**	**$ 177,625.22**	

This document is for informational purposes only. By providing you with this information Egg Geothermal Systems, Inc. is not providing, nor intending to provide, you or any other reader of this document with legal or tax advice. For legal or tax advice you should consult a lawyer or CPA. Neither Egg Geothermal Systems, Inc. nor its affiliates or consultants shall be responsible for your use of this document or for any damages resulting therefrom.

Total Project Amount Qualifying for Federal Tax Credit: $ 76,182.42

Modified Accelerated Cost Recovery System of MACRS on your Investment Example shows Q1 install.

2010 Tax Credit	Total Investment X 10%	$	7,618.24
Depreciable Basis:	Total Investment - (Tax Credit/2)	$	72,373.30
2010 MACRS Tax Benefit	Depreciable Basis X 35% MACRS X 40% tax rate	$	10,132.26
2011 MACRS Tax Benefit	Depreciable Basis X 26% MACRS X 40% tax rate	$	7,526.82
2012 MACRS Tax Benefit	Depreciable Basis X 15.6% MACRS X 40% tax rate	$	4,516.09
2013 MACRS Tax Benefit	Depreciable Basis X 11.01% MACRS X 40% tax rate	$	3,187.32
2014 MACRS Tax Benefit	Depreciable Basis X 11.01% MACRS X 40% tax rate	$	3,187.32
2015 MACRS Tax Benefit	Depreciable Basis X 1.38% MACRS X 40% tax rate	$	399.50
Total MACRS		$	28,949.32

7906 Leo Kidd Ave., Port Richey, FL 34668 Office: 727.848.7545 Fax: 727.848.0134
Lic #: CAC053846, CMC1249626 www.egggeothermal.com

FIGURE 10-5 Commercial geothermal air-conditioning systems often have an ROI that is measured in days or may even be accrued as a net positive from day 1. This is due to U.S. federal tax incentives that use a type of accelerated bonus depreciation known as MACRS. (*Egg Systems*)

Air-Conditioning, and Refrigeration] equipment as a fixed asset, depreciable over 39 years." He went on to say that, "Small business owners need a more realistic recovery period of 15 years to provide an incentive to install high efficiency systems to help them cut their energy costs."

So how does five years sound now?

The Modified Accelerated Cost Recovery System (MACRS) is the methodology in the U.S. tax system for recovery of capitalized costs of depreciable tangible property (other than natural resources). Under this system, the capitalized cost (basis) is recovered over a specified life by annual deductions for depreciation. The amounts are set according to the class of assets and are published in detailed tables by the Internal Revenue Service (IRS). The deduction for depreciation is then computed by the taxpayer via one of two methods, usually by their choice: declining balance switching to straight line or straight line. Figure 10-5 shows the values of MACRS calculations for this example, with a total five-year sum of $28,949.

To combine these incentives, as in Fig. 10-5, we add the tax credit and the five-year MACRS together and come up with a balance of $39,614, which is barely more than a standard high-efficiency system would have cost to begin with. By the time the first-year energy savings are calculated, we are essentially getting the geothermal upgrade for free, plus receiving bonus funds. Without figuring in replacement costs associated with the total ROI, this works out very nicely. Remember that the geothermal equipment has a much longer lifespan than standard equipment sitting outside . . . sometimes up to three life cycles longer.

Net Present Value

Another consideration worth evaluating is *net present value* (NPV). At Egg Geothermal, we have seen overwhelmingly favorable results when the proper parameters are used.

NPV is a finance tool that is often defined as the sum of the present values (PVs) of individual cash flows. Used for capital budgeting, it measures the excess or shortfall of cash flows, in present value terms, once financing charges are met. However, one potential problem when doing a net present value is the lack of understanding with regard to the durable infrastructure.

Let us give an example: If I try to sell you a sink with hot and cold running water for your home, and you have never had running water there before, you would likely be interested. After all, you could save all of those trips to the well to fill your basin or bucket.

So, I tell you the sink and faucet cost $100. Fine, but then you have to get water piped in and a well pump...and you'll need a water heater for the hot water faucet, plumbing for the wastewater...and don't forget a septic tank and a drain field. How much for these other items? Let's see . . . $10,000 for the well, and about $4000 for hot and cold plumbing, and the tank will be $400, plus the electrical for both the well and the tank at $1600. Oh, and the septic tank and drain field is $7000. That's $23,000! Oops, forgot the sink. . . . Make that $23,100.

You might ask me what the payback or ROI is on this project. I would look at you like you're crazy and say: "You either want it or you don't; there's no payback." [2]

There is another big plus: You will now be able to add other plumbing appliances as needed, and if you ever need to replace the sink, you already have the infrastructure in place—you know, the piping and so forth.

The same holds with geothermal: Once the piping is in, it's in. In 30 years, when you replace the geothermal air conditioner (like the sink), you won't have all that added cost.

30 Cents a kWh: The Big Impact of Higher Electric Rates

At Egg Geothermal, we are currently working with several different developers on community geothermal systems. This will help the initial cost on these systems quite a bit. One such project we're working on is in the Bahamas where, as of the time of this writing, the charge for electricity is more than 30 cents per kWh. That is 250% more than the cost in many places in the United States. This means that the payback times there are two-and-a-half times faster than here, without the federal tax credits factored in. So, if we are looking at a payback for a project in the United States of five years before tax credits, the same situation in the Bahamas may be only two years. This example also illustrates the increasing need to decrease our consumption of electricity.

Summary

Geothermal HVAC systems typically pay for themselves in just a few years, at a rate that is quicker than many other renewable technologies, such as solar or wind power. With a few simple calculations, it is possible to estimate an ROI before anything is put in the ground.

[2] Well, at least not in terms of utility bill accounting for your house. There will likely be a significant payback in terms of time savings, which you could then apply toward more productive activities that could make you more money. Also, you will likely have better hygiene, which can reduce health care costs and further boost productivity.

CHAPTER 11

Verifying Your System

My Uncle Bob is my favorite uncle. Almost everyone has an uncle like Uncle Bob. The summer of my 15th birthday, he drove up in a red Ford Bronco to my grandma's house in Downey, California, and asked my cousin Don and me to hop in and go for a ride.[1]

Don was older than me by the margin from July to September. So he was 15 and I was 14 at the time, which he never let me forget. So Uncle Bob let him drive the Bronco. In the 1970s, a full-size Bronco was the highest and nicest four-wheel drive on the street. For some reason, the three-month age difference altogether disqualified me . . . but it's OK, because he made it up later that year, when I turned 15, and was rewarded with a drive in a Ferrari, à la Magnum P.I., down Route 66 toward Barstow. They really do go 160 MPH!

So Uncle Bob took us over to Newport Beach to open a new restaurant he and my Grandpa were starting on the water. We worked there for a while, hauling out the construction trash, earning enough money to spend it all the next day at the local mall, after paying our tithing. As Don and I had the best summer ever, we came up with a lot of concrete assumptions. Among these were the following:

- Uncle Bob's net worth was at least $224 million.
- The restaurant was going to take in between $1600 and $2600 per day.
- The Ford Bronco cost $9000 to $10,000 and had a 400-horsepower engine.
- A wide receiver on a pro football team could run about 35 miles per hour (Uncle Bob told us that one).
- Mercedes Benz was deed restricted as the only brand of car allowed to be driven by residents of Newport Beach.
- "My Sharona" was the coolest song in history.

[1] Don was my favorite cousin. Everything I ever did wrong (on purpose) I can blame on him. It wasn't all that bad, Mom.

Now the facts:

- I'm pretty sure that Uncle Bob was not worth that much, though I don't know for sure . . .
- The restaurant was bankrupt before Christmas.
- The Ford Bronco was $5566 fully loaded, and the 400-cubic-inch engine had 158 horsepower at 4000 rpm.
- A wide receiver in the NFL has never run faster than 26 miles per hour.
- I still don't know about the Mercedes thing. I can honestly tell you that about 9 of 10 cars on the street were Mercedes. . . . Weird.
- Sharona Alperin, who dated the Knack's lead singer, is a real estate agent in L.A. now, and apparently Queen's "Bohemian Rhapsody" is the coolest song ever.

The purpose of this recollection is that we human beings love to believe amazing and unbelievable things. So, how do we know that a geothermal system will perform the way a salesperson promises, once it is in place in the real world?

Actual SEER and EER Results

Take the gentleman who recently had a 29 EER, two-speed geothermal air conditioner installed on a lake loop in south Florida. Now his EER was not actually 29, it was 15.5 EER running at full load, and 18.5 EER running at part load. That 29 EER is correct in certain situations, just as a 19 SEER is attainable in certain situations for an air-cooled unit. That efficiency is a seasonal average and is flawed when there are significantly more heating or cooling hours than average.

The rating systems are a benchmark under which all equipment is tested. It is rarely the best of both worlds. You can, however, be fairly certain that the ratings you see will be a fair comparison one to another, all things being equal.

As discussed in Chapter 7, the efficiency of air-conditioning systems can be rated several ways, but most often it is by the SEER, which is defined by the Air-Conditioning, Heating, and Refrigeration Institute (AHRI) in its standard 210–240 performance rating of unitary and air-conditioning and air-source heat pump equipment.

SEER is related to EER, which is a ratio of output cooling capacity in Btu/h (British thermal units/hour, sometimes written as Btuh) at a given point (see Chap. 6 for more calculations). Coefficient of performance (COP) is often used to reference heating performance of geothermal heat pumps, but can be interchangeably used and correlated with EER by dividing EER by 3.413 (1 watt is equal to 3.413 Btu/h). One more thing: In air-source heat pumps, the normal rating is the heating season performance factor (HSPF), which is an average much like the SEER rating. A HSPF of 10 can be divided by 3.413 to give a COP of 2.9. Air-source systems are normally rated in SEER and HSPF because air temperatures drastically change and need to be averaged throughout an entire season. Geothermal or ground-source heat pumps are normally rated in EER and COP, which are EERs measured at a single point in time. Figure 11.1 shows how efficiency ratings tend to be more stable (and higher) for earth-coupled than for air-source HVAC systems through a typical year.

In Chapter 5, we introduced the standards maintained by the AHRI. Let's review those here, with some additional thoughts.

- *AHRI-320*: Not a geothermal standard, but refers to closed-loop, water-source HVAC units that work with a boiler and

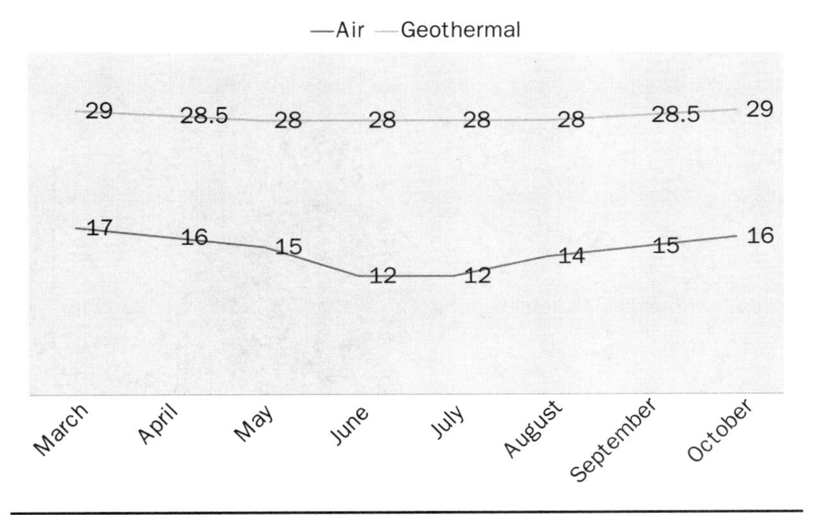

FIGURE 11-1 Seasonal SEER ratings fluctuations: typical example. This chart compares the stable ratings of EER versus SEER; efficiency ratings go down substantially for air-cooled equipment due to extremes of outdoor temperatures. But the EER for geothermal air-conditioning systems is relatively unaffected because they are receiving most of their energy from the more stable and neutral temperature of the earth. (*Egg Systems*)

cooling tower (see Fig. 11.2). In this case, performance numbers do not include the required pumping energy costs to supply the units with water or fluid, or the fan power of the cooling tower. In effect, it gives you false performance figures because there are hidden extra operating costs you don't see. This would be like obtaining a miles-per-gallon rating on a car that is tailgating a semitrailer during the course of the test. The mileage would be significantly improved because of the dramatic reduction of wind resistance, just as the AHRI-320 rating does not take into account the motors and fans in the cooling tower.

- *AHRI-325(70)*: This is for open-loop geothermal systems where average groundwater temperatures vary around 70°F (predominantly the south). These ratings do include pumping energy costs to supply the units with water. Because there are no hidden operating costs under AHRI-325, it evens the playing field for all of the geothermal air-conditioning systems rated by AHRI, as long as they are operating at the prescribed test conditions. But it is up to the design professional, the contractor, and the homeowner to decide what is going to be the best fit for their particular installation.

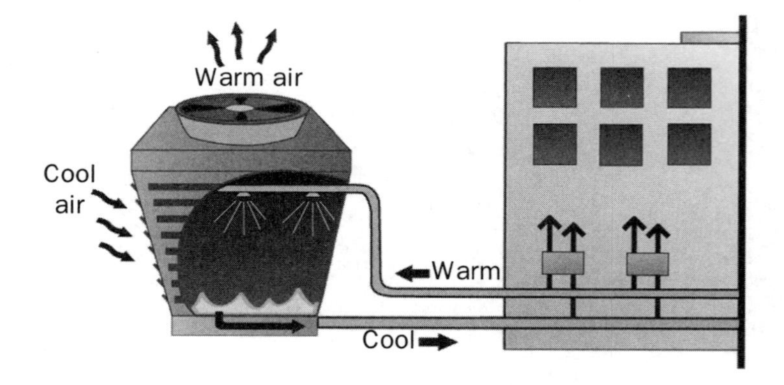

FIGURE 11-2 Water-source equipment used in conjunction with a cooling tower and a boiler is rated on AHRI-320. The efficiency of equipment in AHRI-320 does not take the power used by the cooling tower and boiler into account, making it a flawed rating. In comparing geothermal (earth-coupled) equipment, AHRI-330 is used for closed loops, and AHRI-325 is used for open loops. Both of these ratings take the pumping power needed for the coolant water into account, which means that nothing is missed in the calculations. (*Sarah Cheney/Egg Systems*)

- *AHRI-325(50)*: This is for open-loop geothermal systems where average groundwater temperatures vary around 50°F (predominantly in the north). These performance figures also include pumping energy costs. As you might have gathered, the EER for cooling on the unit rated under AHRI-325(50) is going to be very good. But the heat will be rated for a lower COP than with the AHRI-70 because there is not as much heat in the water to start with.
- *AHRI-330*: This is for geothermal equipment operated on a closed loop. The figures also include pumping energy costs, this time to supply an exchange fluid, typically antifreeze. The standard input for heating is 32°F, and for cooling is 77°F entering fluid temperature. This comes close to giving us a fair starting point to compare one system to another. But remember that you're comparing 50-degree water temperature on the AHRI-325 to 32 degrees on AHRI-330.

In central Florida, it would be typical for a person to have a 5-ton, 10 SEER air conditioner for a 2500-square-foot house. It would also be normal to spend about $2144 annually on the electricity to cool that house. A geothermal air-conditioning contractor might come in and sell this individual on a 5-ton, 30 EER geothermal air conditioner. He would be able to state correctly that the system will drop the cooling season costs down to $692 per year. What an amazing savings! But if that contractor does not install the unit under the conditions for which he just gave the efficiency rating, that will not happen. And indeed, in central Florida, that EER is not actually possible without some help from load sharing or other means.

If you take a look at the actual rating from the ClimateMaster TT series sales literature, as excerpted in Fig. 11.3, you will notice that the 5-ton, model TT-064 has an EER of 29.7. That is basically an EER of 30, and that is what most customers will be told. Here is the key fact: That EER of 29.7 came out of the column for ratings on open-loop systems at a national average of 59°F. The temperature of groundwater in Central Florida is about 72°F. At this temperature the best achievable efficiency is about 24 EER. By the way, that is equivalent to a 28 SEER unit. Before you try to compare that to some of the 19 and 20 SEER standard air-conditioning systems out there, note that *their* actual performance is often lower.

Remember, in our stated example, we got to 24 EER with a pump and reinjection well system. Often times, however, the contractor will talk the homeowner into a closed-loop system. Experience and testing show the resulting incoming water temperature drifting up in the 90 degree range, placing the EER down around 17 EER or lower.

Ground Water Heat Pump				Ground Loop Heat Pump			
Cooling 59°F [15°C]		Heating 50°F [10°C]		Cooling 77°F [25°C]		Heating 32°F [0°C]	
Capacity	EER	Capacity	COP	Capacity	EER	Capacity	COP
Btuh [kW]	Btuh/W [W/W]	Btuh [kW]		Btuh [kW]	Btuh/W [W/W]	Btuh [kW]	
22,200 [6.51]	30.8 [9.0]	18,600 [5.45]	5.1	21,300 [6.24]	26.0 [7.6]	16,500 [4.83]	4.6
30,200 [8.85]	31.5 [9.2]	24,800 [7.27]	5.1	28,900 [8.47]	27.0 [7.9]	22,100 [6.48]	4.5
40,700 [11.93]	28.7 [8.4]	35,400 [10.38]	5.1	39,600 [11.61]	24.9 [7.3]	31,200 [9.14]	4.6
51,900 [15.21]	29.7 [8.7]	49,800 [12.25]	4.7	49,800 [14.60]	25.3 [7.4]	37,500 [10.99]	4.3
59,800 [17.53]	24.5 [7.2]	51,700 [15.15]	4.3	57,700 [16.91]	21.4 [6.3]	45,400 [13.31]	3.9

FIGURE 11-3 Much like an automobile's miles per gallon (MPG) rating, the actual EER or SEER rating of equipment is dependent on a number of factors of use, such as regional earth temperatures and soil types, style of earth-coupled installation, and so on. (*ClimateMaster*)

Factors That Can Affect Efficiency

So the extraordinary energy savings promised at first can become an actual energy saving of $953, a reduction in yearly savings of a little more than $300. But how are you going to know that the system you are told will be best for your climate will work well and at the efficiency you have been quoted? There are several precautions to take:

- Know what types of systems do and don't work well in your area.
- Become educated on what brands of equipment work the best and have the best service available.
- Know how to read the AHRI data included in the sales figures of the equipment and compare it with the stated efficiency and known factors for soil and water conditions in your area.
- Check references on the contractor that you are using.
- Hire an engineer qualified in geothermal HVAC design to at least review the proposal; this may seem expensive, but it's not that bad, and it can save a lot of heartache.
- Follow up with your very own energy efficiency audit. You and your certified technician can do it together, and you'll be able to sleep better knowing how good your system is truly operating.

Regional Climate Issues

This book contains solutions to many of these concerns and questions, to help start you on the right path toward a successful installation.

Here's another example of a flawed proposal in a heating-dominant climate:

In Minnesota, the heating bill for the winter in a 2500-square-foot home averages $2376. By switching to geothermal heating, the specs on a TT060 say that you can achieve up to a 5.7 COP. A geothermal contractor can truthfully calculate this cost at $627 per year to run the system. Quite a savings! But the real data for Minnesota suggest that a closed-loop system is going to experience subfreezing incoming temperatures. Referring back to our AHRI data reduces the heating efficiency to a 4.3 COP. This changes the operating cost to $711 per year. This can go substantially higher if backup heat is needed, as it is normally due to the geothermal source becoming exhausted.

Most thermostats or controllers will automatically pull in a backup source, which is commonly gas or electric heat. If there are no warning signs, these conditions can be present for a month or more before the customer is made aware. Sometimes this is due to utility averaging, wherein the electric or gas utility skips a month or more, averaging usage from years past. Then a customer can be surprised by a gas or electric charge that can be many times what they are used to receiving.

In Florida, 2010 saw an unusually cold winter, accompanied with many sad stories of exceptionally high energy bills, many in the low thousands of dollars. Many of these were the result of heat pump malfunctions that simply were not noticed, despite the fact that thermostats have warning messages or lights to let us know when there's a problem.

How to Calculate Your Own EER

For this procedure, you should have a qualified HVAC professional at your home or business. He or she will take several data points for you, which you can easily convert into your EER with some basic arithmetic. This calculation can be performed on a split system or packaged systems.

The first thing to be determined by your HVAC professional is if the air-conditioning and heating systems are operating up to the specifications for which they were designed. This involves many variables and is common knowledge to HVAC professionals. Once you're assured of maximum performance of your existing or new system, you're ready to collect the necessary information.

Data Points

Collect the volts alternating current (VAC) from the terminal block coming into the unit, as in Figs. 11-4 and 11-5. This should be done while the system is operating at full power. Record it where it says "VAC" in Figure 6-3 (back in the section where we introduced these calculations).

FIGURE 11-4 Much like the ability to calculate your own MPG in an automobile, you can have your service professional take a few readings for you, and through simple arithmetic, you will have your personal EER and COP ratings. Remember that you must use a qualified professional to obtain these readings from your equipment. (*Photo by Egg Systems*)

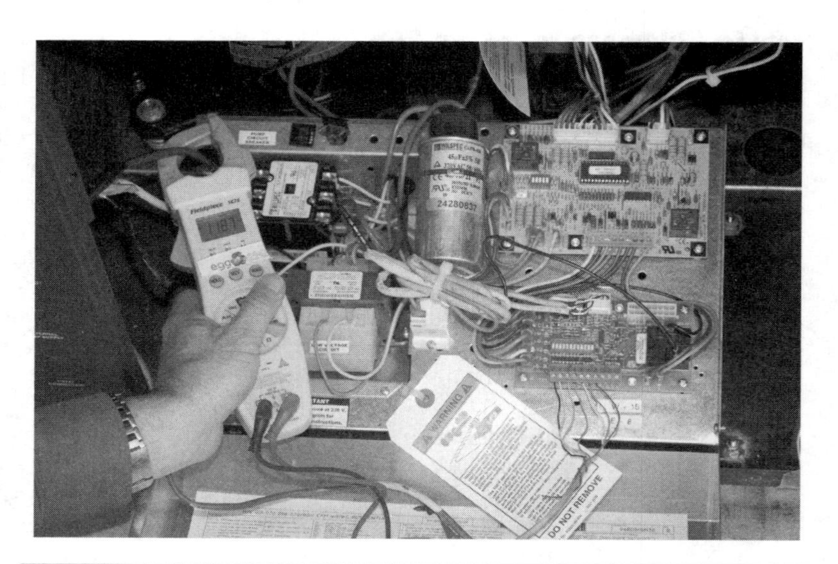

FIGURE 11-5 A technician will use an electronic meter to take the data necessary to calculate the energy consumption of your geothermal air conditioner. (*Photo by Egg Systems*)

At the same location, collect the amperage coming through the main conductors into the unit (Fig. 11-6). Record this in the space marked "AAC" (amperage alternating current) in Figure 6-3.

If your circulating pumps on a closed-loop system are powered from the geothermal heat pump, you are complete, and you have your part load and full load EER rating. If you have a pump and reinjection system, or any other type of external components drawing energy on your HVAC system, this is the point at which you identify those devices and calculate the consumption in watts, just like above. The Btu/h stays the same, but the consumption is increased.

If this unit is a two-speed or two-stage unit, you may drop the load down to its part load capacity at this point and collect the part load amperage at the same data point and record it in "AAC Part Load." Now, you can multiply the two numbers and determine the system's watts. Watts are a true measure of power consumption. In the example we're using, the watts are 3600. Now you can divide that number into the Btu/h rating, which in our example is a 3-ton air conditioner, rated for 36,000 Btu/h. The answer, conveniently, is 10, as in 10 EER.

Since 10 EER was the minimum standard for much of the last 20 years (up until January 2006 in the United States), you will find that most existing systems will be at a 10 EER or less, often down around 5 or 6 EER. You will also find that you can expect up to three to six

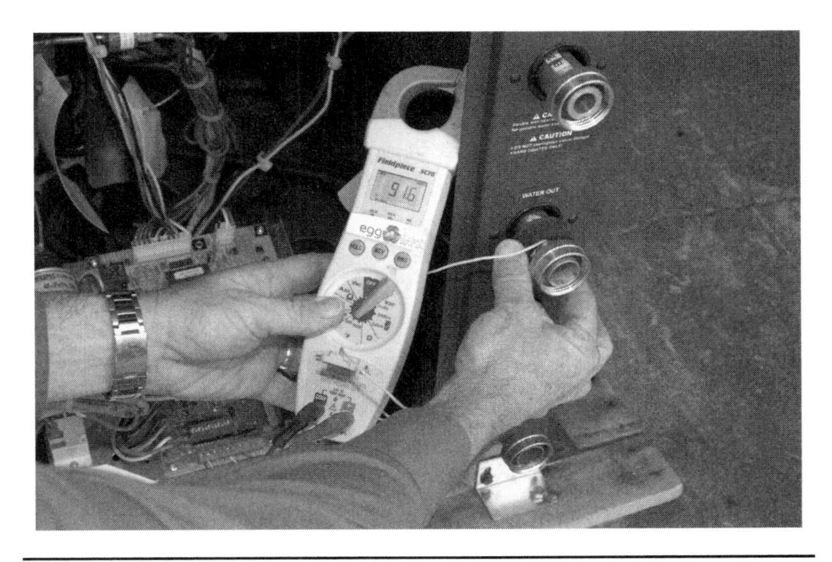

FIGURE 11-6 Other data points such as pressure and temperature of incoming and outgoing earth-coupled lines come into play. (*Photo by Egg Systems*)

times the efficiency by changing to a top-of-the-line geothermal heat pump.

One more fun calculation that puts geothermal air-conditioning further out front is this: EER is not equal to SEER. In order to justify the numbers for residential, you can take the EER rating and divide by .875 to give a corrected SEER rating. For example, even a poorly rated geothermal heat pump running with an EER of 20 is equivalent to a 23 SEER air-source heat pump (SEER = EER/.875). Putting this in perspective, the very best in efficiency for air-source heat pumps is just about where geothermal heat pumps pick up.

Minimum Efficiency Standards

Below is a table of the progression of minimum efficiency standards in the heating and air-conditioning industry over the next few years. In order to understand it, we need to introduce the concept of the *degree day* (DD), a measure of heating or cooling. A degree day is technically an integral of a function of time and temperature. They are defined relative to an appropriate base temperature (typically 65°F or 60°F), which is the outside temperature that would require no conditioning. One popular approximation method is to take the average temperature on a day and subtract that from the base temperature. If the value is less than or equal to zero, that day has zero DD.[2]

Degree days can be added over periods of time to provide an estimate of seasonal heating needs. For example, the degree days in a heating season in New York City average 5050, whereas they are 19,990 for Barrow, Alaska. This means that a home of similar size and design would require roughly four times the energy to heat in Barrow versus New York. In contrast, Los Angeles has degree days of 2020.

Regional Minimum Efficiency Standards

- 14 SEER in <5000 DD Climates (12.2 EER in West*) in 2015
- 90% AFUE in > 5000 DD Climates in 2013
- Agreement between AHRI, ACEEE, ASE, CEC

Building Codes (new IECC Baseline by 2013)

- 15 SEER in <5000 DD Climates (12.5 EER in West)
- 14 SEER in >5000 DD Climates (15 SEER HP)
- 90% AFUE everywhere (92% in >5000 DD and West)

* Equivalent of just over 14 SEER
AFUE = Annual Fuel Utilization Efficiency

[2] This is often written as HDD, for heating degree days.

FIGURE 11-7 As with all HVAC systems, once you have an overall understanding of the data points, you can have a productive and informative conversation with your geothermal professional about your needs. (*Photo by Egg Systems*)

Summary

We have seen that it is critical to verify that your installed system is working as intended. Luckily, the process is relatively straightforward, as outlined in this chapter. There are a number of factors that can affect efficiency, such as regional climate issues, thermal retention, and improper installation, as well as other problems with your equipment.

CHAPTER 12

Life Cycles and Longevity

I go to a dermatologist every six months, or more often as needed. This is not just a preventative measure, because every six months I need to have several, sometimes a dozen or more, little precancerous spots frozen on my face and body. Twice, I've undergone full-blown cancer removal and plastic reconstruction. There is a remarkable new procedure wherein a chemical is applied to my face; then the affected area is bombarded with UV rays for about 30 min. It's painful, but when it peels, new cancer-free skin is born.

Machinery, unfortunately, can't be reborn as easily. The worst thing for a piece of equipment is exposure to outside temperature and weather extremes, which cause damage through oxidation (rust) and weathering. I have an 18-speed mountain bike that I've used for the last 23 years, and it still runs flawlessly. The bicycle has normally been kept in a garage, but it sat in the rack in my front yard for a month last summer. It aged more in that brief time than over the past two decades. I thought it was not going to be recoverable. But after an hour of restorations, and about five ounces of WD-40, it was back in working order.

Similarly, an average commercial rooftop air conditioner gets no relief from the elements. There is typically no shade, plenty of wind and rain, and no one to wash, wax, and maintain it under normal circumstances. It is, therefore, not so surprising to see a 10- or 12-year-old piece of equipment that is quite literally falling apart.

On the opposite end of the spectrum, it is not uncommon to see a classic car, such as a 1968 Mustang, in beautiful shape. The reason is simple: It has been kept inside.

The Benefits of Indoor Equipment

Geothermal air conditioners remove that outside equipment variable in a way that no other system is able to do. And there are more benefits of this than just antiaging. Here are a few other reasons why outdoor equipment is at a disadvantage:

- It is susceptible to damage from lawnmowers and weed eaters.
- It can be cross-contaminated by other exhaust sources, such as dryer vents.
- It can attract bugs, rodents, lizards, and other pests that can also cause equipment problems.
- It can be a danger to pets and children.
- It can cause noise pollution.
- It can be susceptible to vandalism and theft.

In contrast, my company services a geothermal system in a 3000-square-foot home in Clearwater Beach, Florida, that works great after more than 30 years. It was originally a directexpansion geothermal AC system, in which each room had a valence, behind which was an evaporator pipe with fins. This had a piece of PVC pipe cut in half the long way to catch the drips of condensation from the fins and channel that outside the home. It used no fan indoors or out, but had a coil of half-inch copper run into seawater under a dock. Over the years, we've replaced the geothermal compressor sections and upgraded the loop to a polyethylene closed loop. But these upgrades have been relatively minor, and the system lives on.

IMPORTANT NOTE: *In our experience, geothermal HVAC equipment can readily last as long as three life cycles of conventional heating and cooling systems (Fig. 12-1), yet geothermal systems require a third of the maintenance (Fig. 12-2).*

How to Determine When Upgrades Pay Off

Typically, replacement of a piece of HVAC equipment after 8 to 12 years of service will result in a substantially lower electrical bill. This is because efficiency advances are continually being made and because outdoor equipment faces such a fantastic beating by the elements that it can't hold on to its efficiency rating too long.

Air-conditioning equipment usually has a five-year compressor warranty, although some brands have 10 years. Often, the heat exchange coils will also have a five-year warranty, if not used commercially, and the electrical components, such as fans, motors, boards, and relays, will have a year or two.

So here is how the life cycle recommendation boils down on air-source (nongeothermal) equipment: If the equipment is under five

FIGURE 12-1 Two new AC condensers beside two 9-year-old condensers, showing how quickly they age in the hot sun and cold rain. The older units use twice as much energy, although they are all rated at 10 EER. (*Photo by Egg Systems*)

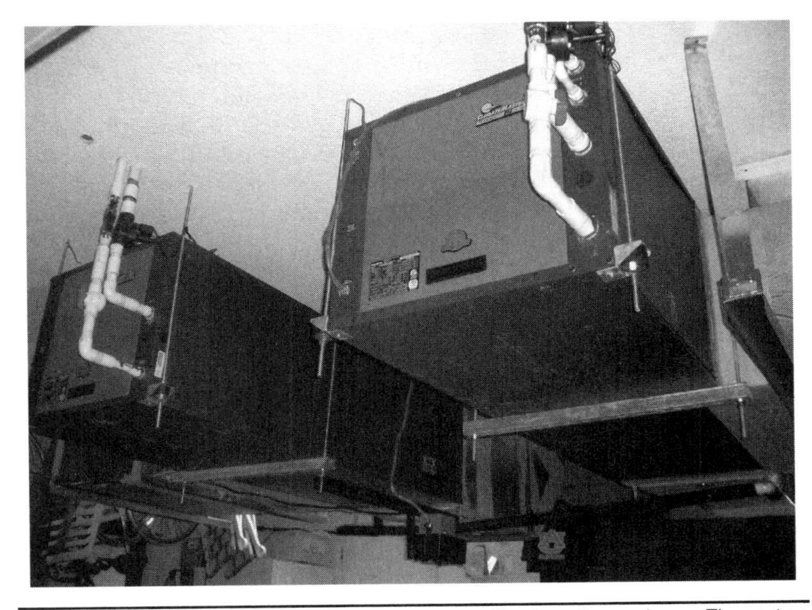

FIGURE 12-2 Geothermal air conditioners can fit almost anywhere. These two units were moved into a garage rather than the attic to take advantage of the more even temperature and humidity. (*Photo by Egg Systems*)

years old, there is usually no reason to recommend replacement or upgrade. If the equipment is more than five years old and it is not warranted for 10 years, and a compressor or a heat exchange coil is faulty, it may be time to consider an upgrade. This is because the cost of a new compressor is often 70% or more of the cost of a new conventional system. A replacement compressor may only be warranted for 90 days, while a new system usually gets five years.

If a fan motor or other semiexpensive part fails at 8 to 12 years, it is often advisable to upgrade because of the likelihood of beginning to pour hundreds or thousands of dollars into an air conditioner that is arguably on its way out of commission.

If the air conditioner is in need of any repairs after 10 years of age, it is usually recommended for replacement. This is partly due to the liability of repairing the unit, and then getting called out a few months later when a more expensive part fails, and getting an earful along the lines of, "Why didn't you tell us to just replace it back then and I would have saved the service call?"

At 14 years of age, many HVAC companies will all but refuse to work on the system, referring to it as a liability.[1]

A new geothermal air conditioner by itself is not too much more expensive than a high-efficiency standard system. So, once the loops are installed, replacement costs are very close between geothermal and standard high-efficiency equipment (Figs. 12-3 and 12-4). With this cleared up, there is only the maintenance aspect of the two systems to compare.

Historically, the most common equipment failure in a standard residential system is the condenser fan motors, followed closely by contactors (electrical switches that operate the different phases of the systems) and compressors, all of which are normally outdoors and subjected to harsh conditions. Geothermal air conditioners do not have condenser fan motors. They do have compressors and contactors, but properly installed, they operate at much lower loads due to the indoor location of the equipment. This helps the parts last longer and have fewer problems (Fig. 12-5). It also helps explain why geothermal systems tend to require a third less maintenance over time than conventional HVAC options. That, of course, means less money spent on service calls and less hassle.

[1] It is true that heating furnaces, which are located indoors, will often last 20 to 30 years and we regularly run into functioning 40- and 50-year-old oilburner furnaces.

FIGURE 12-3 This customer elected to stay with the option of outside geothermal condensers, so he built a roof over the top to protect them from the elements. They were installed in 1992 and still work great. (*Photo by Egg Systems*)

FIGURE 12-4 This Florida home is serviced by the condensers in Fig. 12-3. The project was one of many made possible by grants from the U.S. Department of Energy, with assistance from the University of South Florida (and Egg Geothermal). (*Photo by Egg Systems*)

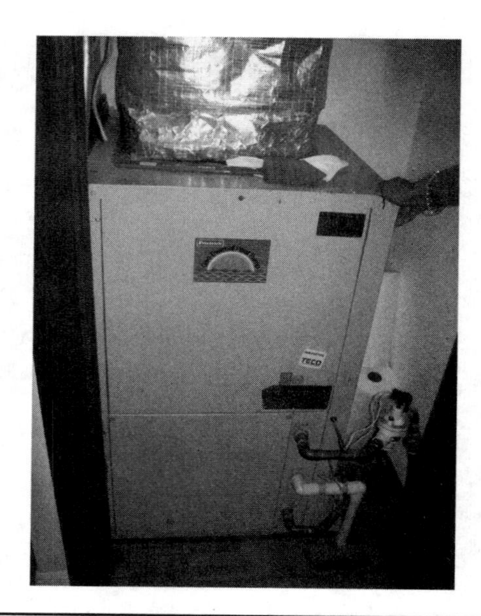

Figure 12-5 This Friedrich geothermal system was installed in 1978 to serve a 2000-square-foot house in Florida, and it didn't require a service call for any repairs until 2010. Friedrich became ClimateMaster in 1985. (*Photo by Egg Systems*)

Summary

The Oregon Institute of Technology has done life cycle cost analyses of geothermal systems and found that although initial costs are a bit higher than conventional alternatives, everything else is dramatically lower for geothermal from energy bills to maintenance and repair.

To summarize the life cycle cost effectiveness of geothermal air-conditioning:

- A geothermal system that is properly engineered and installed is expected to cost about one third as much to maintain as standard air-conditioning.
- A geothermal HVAC system is expected to last roughly three times longer than a standard system.
- A geothermal system is easily upgradeable and expandable to other appliances, such as domestic hot water and pool heating, with the proper preengineering.

CHAPTER 13

Common Problems and Horror Stories

Experience: that most brutal of teachers. But you learn, my God do you learn.

—C.S. Lewis

We all have memories of experiences we wish we could share with the whole world . . . so that everyone did not have to go through the same dumb mistakes over and over again. As George Santayana wrote in his book *The Life of Reason*, "Those who cannot remember the past are condemned to repeat it." I feel a little validated here in that many of my mistakes were truly first generation. The problem now is that many of us who are veterans are seeing terrible repeats of history that simply are not necessary. That is why it is ultimately up to the consumer to be educated. But contractors must shelve their pride long enough to realize that it is OK to ask questions, and make the call for a second opinion. We're all here to help.

Early on in this book I told the story of how I thought I had discovered geothermal cooling with my well and water-to-refrigerant exchanger. That home in Palm Harbor, Florida, was upgraded a little later on to a state-of-the-art Addison geothermal air conditioner. It is still operating very efficiently on a pump and sprinkler system. My confession on that particular job involves the shallow well. In most of Florida, the water level is fairly high. There is often a layer of limestone beneath the sandy surface trapping water in what is called the *surficial aquifer*. Many irrigation wells in Florida pull from this shallow aquifer, providing a seemingly unending source of fresh water.

Then we had a drought year. I caught endless disparaging remarks from neighbors and water authorities about the sprinklers that were running at all hours of the day. Finally, I placed a sign prominently on the lawn stating, "Using reclaimed water."

But then the well ran dry. Granted, it was just 30 feet deep and jetted right into the sand. I had to call out the well driller and spend $4000 to drill 150 feet into the Floridan aquifer. This is still in operation today, and will likely be working for a long time to come. The lesson learned is that we now recommend deep wells.

A Word on Water Conservation

Just because something doesn't seem that expensive is not a reason to waste it. Once again, this is a big part of Provident Living, taught in church. I remember well the little rhyme placed on the wall in my home, as I grew up the oldest of nine kids: "Use it up, wear it out, make it do, or do without."

To me, renewable means that we reuse whatever we can, or the resources used are returned without any permanent degradation. So to be renewable-wise is to be responsible with resources, such as water in the case of geothermal HVAC. Just because we can pump water legally for use in a geothermal air conditioner, doesn't mean we should not put the water right back where we got it.

Common Issues

There are many things that can go wrong when it comes to installing an earth-coupled system. Oftentimes, honorable contractors will absorb much of the resulting costs. At Egg Systems, we like to take the "insurance company" type of approach. We charge a fair price and stick to it. If we lose money too many times due to unforeseen situations, then we'll adjust our rates for higher exposure or losses.

Underground Hazards

In some places, geothermal installers have unearthed debris from past construction, old junkyards, or other activities. In such a case, sharp and jagged edges can kink and bend the polyethylene loop pipe as in Fig. 13-1, causing leaks and restrictions that render the system inoperable. Dealing with this properly will cost a bit more, but it is worth the effort.

The solution—if obstacles can't be easily and completely removed—is to bring in sand to cover the debris before backfilling. Sand protects and cushions the bed, while giving a great medium for exchanging temperature with the earth.

FIGURE 13-1 It is important to be aware that variations in soil types are ever present, and care should be exercised to prevent kinking and cutting of the piping during back fill. As an example, sand may be recommended to aid in the backfill process in certain situations, both to protect the piping and boost the thermal conductivity of the overall loop exchanger. (*Sarah Cheney/Egg Systems*)

Pressurized Pockets

Another potential problem facing contractors is the presence of naturally pressurized water and gas pockets. One contractor told us that when the driller hit one of these pockets, what followed was a temporary version of an artesian well, which spilled so much water on the thawing earth that it became an impassable mud bog.

Since they had already decommissioned the system that was heating the home, they had no choice but to have rock hauled in to create a road and surface suitable to finish the drilling for the geothermal well installations.

Broken or Damaged Loops

If a loop is damaged for any reason, it will leak out the water-antifreeze solution, rendering the geothermal air conditioner inoperable. Like any system, down time and availability of technical repair personnel varies, although it is important to point out that geothermal systems tend to be considerably more reliable than conventional options.

There have been instances in which we have employed a temporary measure for down systems. This involves coupling the city water supply to the purge valves and opening the discharge enough to provide ample flow to keep the building cool or warm (see Fig. 13-2). This quick fix is wasteful and jacks up the client's water bill, but is sometimes necessary in a pinch. In these cases, it is vital that the loop be repaired and placed back into service as quickly as possible.

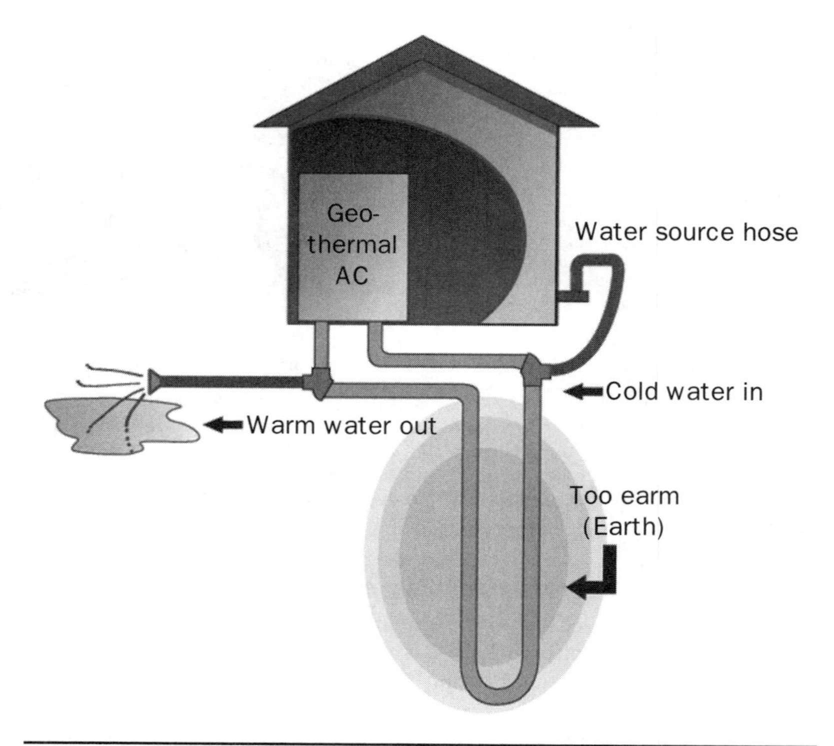

FIGURE 13-2 If an undersized earth loop is unable to handle the air-conditioning or heating load, "leaching" of the system may become necessary as a temporary measure. In such a case, the system will probably need some reengineering. (*Egg Systems*)

Design and Installation Fails

Now, as promised, we'd like to share a few of the disasters that you will be able to avoid by taking the necessary precautions. As you have hopefully garnered from this book, the industry is unfortunately plagued with subpar jobs. But that seems to be changing, and we hope to arm you with the tools to make sure everything goes according to plans.

Choosing the Wrong Loop Type

As we mentioned previously, when Egg Systems installed a geothermal HVAC system in a home in Tarpon Springs, Florida, in 1992, we went with a closed slinky mat loop, buried seven to eight feet in sugar sand, below the mean surficial water level. Although we oversized the system by 12%, and although a slinky loop is a proven technique with thousands of successful installations around the world, it turned out that this wasn't the best choice for this job.

The homeowner started to see problems with the system by the end of his first summer. His incoming water temperatures were above normal, but we were hopeful that the loop field would rest and dissipate the heat over the winter. Well, the problem got worse, until around the end of the third summer, when the slinky mat loop killed the lawn with its excess heat. If I had not seen it myself, I would not have believed it.

We ended up having to supplement the loop with an additional pump and reinjection system. I offered to switch it completely to an expanded open-loop system, but this customer was so proud of his original loop that he wanted to preserve it at any cost.

The problem here was that this closed-loop system was not ideal for this customer's soil and climate conditions. As Dan Ellis, president of Oklahoma City-based ClimateMaster, warned in Chapter 4 with some commercial situations, too much heat retention was taking place. This is just one of many poor design decisions that can be made.

Unfinished Jobs

At a geothermal distributors' "state of the industry" address in Maryland in spring 2010, Dan Ellis of ClimateMaster spoke about faulty installations. Without naming names, he mentioned a high profile home that was in the process of having a geothermal system installed, but with some problems.

According to Ellis, the contractor had bid the price and assurances had been made of a timely completion. The interior HVAC technicians began work about the same time the driller started boring. The previous heating system was removed and the new geothermal air conditioners were installed and ready to be connected to the vertical loop system. That was just about the time that the drilling contractor pulled off of the job, allegedly because the ground conditions were not as he had expected.

Some contractors seem to have no problem quitting a job and leaving a customer and other contractors with no recourse. Ellis said this home was without heat for weeks before the job could be finished. This type of problem is far too common in the industry. And, unfortunately, it happens at the worst times. Oftentimes, this is due to uneducated consumers as well as contractors.

Once again, please have your system engineered by a reputable firm with lots of experience and have it installed by a contractor and driller with a proven track record. And by the way, a well-planned project will close the loopholes on a contractor who might try to pull off the job. This is done through several means, especially the use of bid bonds and performance bonds. These instruments show financial strength, and they ensure the project will be completed with a financially binding agreement that no contractor can afford to lose on. They especially should be considered for larger projects, where there is more to lose. Performance bonds are financial guarantees that are the equivalent of the contractor placing personal funds in the bank for the sole purpose of guaranteeing a satisfactory job. If the job fails to meet the contractual obligation, the funds are eligible for distribution to complete the job.

Pushy Contractors

Another scenario that is tough to deal with for most anyone, especially the elderly, is the overbearing contractor or salesperson. I have talked to homeowners who really wanted to choose the system our company proposed, but went with another outfit due to an intimidation factor. Most often, this follows a barrage of expletives and ridicule, such as, "How could you possibly be so stupid as to go with them? The last 14 systems they put in have ruptured and caused hundreds of thousands of dollars in damage, contaminated the earth for 16 square miles, and caused a near fatality for an elderly woman."

You would be surprised how often this tactic is used, despite baseless accusations. If you hear something like this, just call the authorities and have a restraining order placed on the salespeople who are bothering you. Really. Almost every market has one or two, and they eventually end up on the 6:00 news on an exposé of crooked operators.

Miscommunication, Faulty Equipment, Inexperience—Oh My!

I recently met with a gentleman who is a successful developer in the Bahamas. After a short conversation, I agreed to come take a look at a project he was working on there. So I flew to the islands, and was then driven about 30 miles in a cramped car, along a route that reminded me of the beginning of *Jurassic Park.*

We arrived at a 500-acre site that was bustling with activity and earth-moving equipment, as the landscape was being reshaped for a new town, complete with farms, homes, businesses, schools, and other facilities. But no room was made for gasoline-powered cars or anything else that might make noise, other than the music of nature, provided by waves crashing on the beach and wind whooshing through palm trees. The homes were built up high, with open plans, to take advantage of the breezes and the gorgeous views.

However, there had been problems in paradise. The first was that cars had been gradually let back in, despite the stated intentions of the community. Somehow, there had been an exception made for a busy doctor, then another for the firefighter, then for the high-powered executive. The second problem was air conditioners. They had never

FIGURE 13-3 Working in a trench during a geothermal loop installation. (*Photo by Egg Systems*)

seen it coming. After some of the homes were completed, the sounds of nature were drowned out by the incessant cycling of air-conditioning condensers. So that was why I was called out to this remote location: Geothermal HVAC systems do not need outside condensers, and are essentially silent.

When I returned to Florida, I met with the developer's partner, an engineer who, it quickly became apparent, had had a cacophony of catastrophic experiences with geothermal systems, spanning New York, the Florida Panhandle, Maryland, Idaho, and Nassau. Each of these had been a labor of love for him.

For the New York project, he said he waited two months for a bid on a geothermal system, but finally agreed to just have the system put in by a local contractor, who, allegedly, admitted he did not know how to do it. So after the equipment and ductwork were installed, this developer hired a boilermaker to work side by side with him on engineering the project on a "learn as you go" type of approach. He told me this was time-consuming, frustrating, expensive, and, ultimately, a bad idea. He continues to have problems with the water flow and injection on this system.

The home in the Panhandle apparently has a geothermal split system. Part way through the work, the contractors allegedly told him it needed to be changed over to a unit with two-speed compressors. Even after making the suggested changes, he says it fails to cool the upstairs well. On the home in Maryland, he realized that the first contractor's blueprints were a disaster. So he brought in another contractor, but claims they aren't doing much better.

This developer also claimed to have had bad experiences with at least three different manufacturers' brands of equipment, citing problems ranging from control board failures to air coil leak ruptures and refrigerant leaks. He cited a fourth manufacturer who had produced an electronic control board to fit his requirements for a custom home with perhaps 30 fan coils (air handling units). When it was determined that the controls did not perform as intended, he tracked down the engineers who had done the design. They allegedly told him that they had never done anything quite like it before, they wished they hadn't tried, and that they would not support it. Later, he also learned that part of his directions to the manufacturer of the fan coils had been lost in translation somehow.

A qualified engineering firm should have handled these projects. In our conversations, the developer said it had never occurred to him to hire another engineer to help him, even though he didn't have experience doing geothermal systems. Here he was an engineer himself, yet he had trusted the contractors to watch his interests.

The two of us had an enlightening conversation over about five hours, and the result is that Egg Geothermal will be engineering perfectly silent and out-of-sight mechanical infrastructure for the new town in the Bahamas. A host of structures, from homes to businesses, medical facilities, a water treatment plant and more, will have at least 20 years of sound renewable service for sustainable economic growth. Hopefully, many more are on the way (Fig. 13-3).

Misunderstanding the Technology

In the first week of March 2010, *The Washington Post* ran an article by Christopher J. Gearon about his experience getting geothermal air-conditioning installed in his Derwood, Maryland, home. One of the questions posed to Gearon by an online reader was where to find a good, readable book on geothermal HVAC. If the reader was like us, he may have only come across a handful of dated materials and dry technical information. "That's a great idea for my next book project!" Gearon replied.

Some of the other comments on the *Washington Post* article suggested how little the public really knows about this emerging technology—with which, by the way, Gearon turned out to be very pleased. One gentleman wrote, "How is it possible to have 400% efficiency?

You can't create more energy with this system. The electricity being used to run the heat pump will not be even 100% efficient. No system is. That's basic laws-of-thermodynamics-type stuff. It's not just hard to beat—it's totally impossible! It is an efficient system for sure but let's get the science right in the science section!" Other commenters had the same idea.

It's clear that many people just don't understand that a heat pump moves and concentrates available heat. A heat pump does not use the electricity to create new heat, it takes advantage of the ability of the ground to serve as a heat source or sink. As we saw previously, a geothermal system can attain 5 kW of heat with just 1 kW of electricity.

Unfortunately, there are many other misconceptions when it comes to geothermal HVAC, such as that it requires proximity to active volcanoes or hot springs, or that it only works in heating- (or cooling-) dominant climates, or even that it requires a certain soil type. There is much education that needs to be done.

Summary

I was talking recently with one of our in-house consultants in the Tampa office, who has been in the geothermal air-conditioning business since 1982. He worked side by side with the pioneers of the modern industry, and many of his peers are running the companies that manufacture the technologies today. He remembers that the old way to do an installation was essentially to "Put a roll of pipe per ton in the ground and don't come back with any extra."

While my colleague worries that not a lot has changed out there since, I think the world is much better connected, so we know what works well in which area, and what does not work. With a properly engineered design, we can meet the needs of most clients from Alaska to the Bahamas, or from Germany to Greece, as we found by doing interviews for this book. There are a number of potential problems when installing the technology, but there are also many solutions, as we have seen.

CHAPTER 14

Geothermal Spreads around the Globe

Geothermal HVAC continues to gain momentum around the world, with many exciting projects in the works. In Indiana, Ball State University is undertaking a massive effort to replace four aging coal-fired burners with a campus-wide geothermal system, which will heat and cool more than 45 buildings and save the school an estimated $2 million a year in operating costs.

According to a December 2009 report by the United Kingdom's Environment Agency, ground-source heat pumps could produce a third of that country's renewable heat requirements by 2020, helping the British meet their goal of sourcing 15% of total energy from clean sources. Tony Grayling, head of Climate Change and Sustainable Development at the Environment Agency, predicted that there could be 320,000 geothermal systems in the United Kingdom by 2020, installed at a rate of 40,000 units per year.

That's roughly the number of earth-coupled systems installed annually in North America during the years between 1999 and 2005. After that, sales picked up, and surpassed 100,000 a year by the end of 2009. Two big reasons for this were a rise in oil prices and the enactment of a $2000 per geothermal system tax credit in late 2008.

The future looks bright for geothermal HVAC on both sides of the Atlantic. According to a high-growth estimate, the annual installation rate in the United Kingdom could hit 400,000 by 2019, resulting in a grand total of 1.2 million systems in the country by that year.

Interestingly, the UK's Environment Agency has acknowledged that open-loop, pump and reinject technology has a larger thermal output than closed loop, although they have stated that they are unsure of the potential impacts of the procedure. What they are going to find out when they do get a full report is that R-410A refrigerant is benign and inert, and that the oil for the refrigeration system is of a food grade.[1] Additionally, the exchangers

[1] The oil in a refrigeration system is used to lubricate the moving parts and is carried through the system with the refrigerant flow.

through which the water passes are well constructed, and unlikely to allow leakage. When they investigate further they will find that regulatory agencies that have a lot of experience overseeing the technology are nearly 100% in favor of pump and reinject and use permit requirements that lend a great deal of confidence to the technology.

Take a look at this regulation from the South West Florida Water Management District:

> 40D-3.042 Multiple Wells Under a Single Permit.
>
> (b) A Class V air conditioning heat pump system consisting of one supply well and one return well; may be included under one permit provided the conditions of subsection (2) are met.
>
> (2) A multiple well permit as described in subsection (1) will be issued provided:
>
> (a) The wells are constructed in the same geologic material, completed in the same hydro geologic unit, and drilled on a contiguous tract of land owned or controlled by the same individual or entity; and
>
> (b) Each well is the same diameter and constructed of a similar material.
>
> (3) The District requires both a supply well and a return well in the construction of an open heat pump system. A supply well without a return well is not permitted.

And here is the regulation from Sarasota, Florida:

> Heat Exchange Wells
>
> Thermal exchange process wells as described in chapter 62-528.300(1)(e)1, F.A.C. also require a permit from the Department. These are systems that are used for heating/cooling purposes where there is no change in water volume or chemical composition.
>
> These systems fall into two types:
>
> 1. A system where fluid is circulated through a continuous section of buried pipe such that the earth is utilized as a thermal exchange medium, but no fluid is either extracted from or injected into any underground formation. This type of well does not receive a DEP permit. Multiple wells may be submitted on the same permit application.
> 2. A system composed of a supply well and an injection well where water is withdrawn, used for thermal exchange, and then returned to the same permeable zone from which it was removed. Multiple well systems must have both a supply well and injection well. Both wells must be similar in depth, casing depth, and constructed of the same materials. A single permit may be submitted for both wells.

In some places geothermal wells are as easy as pie, no questions asked (Figs. 14-1 and 14-2). At Egg Geothermal we recently dealt with a permit for a 4000-gallon per minute well; that is a lot of water! It is of no real concern when it is all to be returned to the source whence it came.

I recently had an interesting conversation with a man from Pennsylvania, who called Egg Geothermal to have some engineering confirmed on a project. This gentleman was a former mission president for the church and had a couple of PhDs in financial areas, having written a few books on the subject, including one for McGraw-Hill. He had a three-heat-pump geothermal system proposal, and it included several deep closed-loop wells to provide the proper heat transfer needed for his heating and cooling needs. He mentioned that down in his basement, and about 40 yards through some kind of tunnel (sounds very *Goonies* to me!), he had this apparatus that gathered water from a spring from three different spots on his property, and funneled it to an iron pipe, which poured into a little babbling brook that ran through the town. He said he thought it might produce about five gallons per minute of 50-degree water. We typically refer to wells that are free flowing like this as *artesian wells*, and they are also fairly common in Florida.

I was intrigued, and I asked about the difficulty of drilling for water. He said it was typically rather shallow and he thought he could easily have a supply and injection well put in . . . his only real hurdle being the local water authority. I wished him well on that, letting him know that a good system like that would do better for efficiency than his proposed closed loops.

FIGURE 14-1 Western Heights Middle School in Oklahoma City is conditioned by 88 ClimateMaster heat pumps that work on a vertical ground loop made up of 200 wells, each 350 feet deep. The system saves the school serious money. (*Photo by ClimateMaster*)

Figure 14-2 The pump room at Western Heights Middle School. (*Photo by ClimateMaster*)

Geothermal HVAC Efforts Around the World

Although the United States still has the highest number of geothermal HVAC systems installed—600,000 to 1 million—other regions are now experiencing faster growth rates for the technology. European markets now get two to three times more new units annually than the United States, according to a recent report for the DOE by Patrick J. Hughes. Growth rates are also faster in China, South Korea, and Canada.

Let's take a brief survey of geothermal projects around the world.

Australia

Analysts are excited about the potential for geothermal technologies in Australia, and the industry is growing. A number of key buildings there have had earth-coupled systems since the early 1980s, with numerous deployments in Tasmania, Victoria, and New South Wales. The Geoscience Australia in Symonston (a suburb of Canberra) is heated and cooled by 350 wells.

China

According to the National Renewable Energy Laboratory, geothermal heat pumps were introduced to China in the 1990s. In 1997, the U.S. Department of Energy entered a partnership with a Chinese counterpart to help develop the technology there. Several demonstration projects followed, including a system that served more than 500 residential units.

A recent survey in China estimated that sales of geothermal heat pumps there increased at a yearly rate of 17% between 2005 and 2010. Sales for 2010 are expected to top 7.65 billion Yuan—a 116% increase from 2005 sales. This growth is encouraging since China is the world's largest emitter of greenhouse gases and suffers from terrible air quality. The latter problem plagued athletes at the 2008 Beijing Olympics and is estimated to result in some 400,000 premature deaths a year, according to a study by the Chinese Academy on Environmental Planning. China has been adding dirty coal plants at a breakneck clip to feed the engine of its economic growth, but geothermal heating and cooling could help slow this dangerous pace by reducing the need for electricity.

Eastern Europe

Geothermal HVAC systems are gradually spreading across parts of Eastern Europe. Belarus, the country most contaminated by fallout from the Chernobyl nuclear disaster, has been pursuing clean, safe geothermal technology, perhaps as a partial alternative to nuclear power. A geothermal heat pump was installed in the town of Brest in 2007 to supply a local greenhouse complex. Several municipal heat pump systems have served Minsk since the 1990s.

Slovenia is also slowly adding geothermal. By 2007, the country had installed an estimated 1.6 MW of capacity of geothermal pool heat pumps and some 300 ground-source heat pumps, totaling 3.3 MW. Other Eastern European countries are getting into the game as well, particularly Russia, where projects are being evaluated in the Lake Baikal region and elsewhere.

South Korea

The geothermal industry in South Korea is booming with the number of installations roughly doubling every year. The Korea Institute of Geoscience and Mineral Resources is supportive, and analysts are bullish on the market.

Western Europe

As stated previously in this book, Iceland, Norway, and Sweden have the highest per capita use of geothermal systems. There is also considerable development in the United Kingdom, Ireland, Germany, Switzerland, France, Italy, the Netherlands, and elsewhere.

By 2007, Austria had some 25,000 heat pump installations, according to the World Energy Council. Sweden had 275,000 by 2005, including several hundred large-scale systems that service entire heating districts. Denmark has tens of thousands of heat pumps and support from DONG, the Danish energy company, and the Geological Survey of Denmark and Greenland. France's 2004 Energy Law has also helped promote geothermal systems, and it's estimated that by the end of 2010, the country will be adding 40,000 heat pumps a year in residences, especially in the Paris area. The technology is also becoming available in Greece, where Costas Koutsogiannis, an installer based in Athens, told us he has "done a lot of pond and lake systems," as well as horizontal and vertical loops.

There is much to be learned about the progress of this technology around the world (Figs. 14-3 and 14-4).[1]

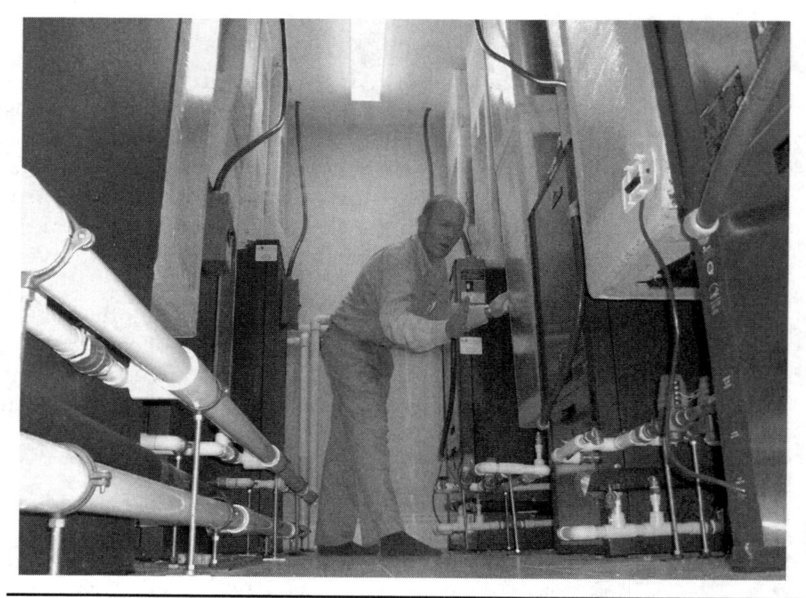

FIGURE 14-3 Multiunit installations offer a number of advantages, in addition to serving larger loads. For example, you can often run both fans but only one compressor much of the time, cycling the second compressor only when there is peak demand. (*Photo by Egg Systems*)

[1] Egg Geothermal keeps a database of developments of the industry from everywhere; please contribute any comments via www.egggeothermal.com.

FIGURE 14-4 At its core, geothermal HVAC is an exceedingly simple concept. It offers numerous benefits and has great potential for wide adoption around the world. (*Photo by Egg Systems*)

Conclusion

Despite impressive growth and success, the geothermal HVAC industry still faces a number of challenges. The public does not really understand us or what we do. This is apparent from comments on news articles as well as a good deal of media coverage itself. For example, I recently saw a TV reporter ask a geothermal contractor if all of that dirt underground didn't make it dusty inside the home. Another question posed was, "What if you want the home warmer than the earth temperature?"

Here are some of the questions that might still be lingering, and that we commonly get:

Is the technology worth it, if it uses some electricity? Yes, a geothermal system takes some electricity to run, but it also decreases fossil fuel consumption considerably. On balance, having a geothermal system results in substantially lower annual utility bills (40–75% less is the typical

Geothermal heat pump: the natural way to heat and cool your home

You can't see it. You can't hear it. Guess you'll just have to "feel" it.

What is it? It's the geothermal heat pump, an energy-saving system that heats and cools your home so efficiently, you don't even know it's there!

You can't see or hear it because the geothermal heat pump does away with your home's outdoor air conditioning unit — you know, the one that sits outside above the ground.

Instead, the geothermal heat pump is part of the home's indoor air handler. Connected to it is special polyethylene piping that's fused together to make it air tight. The piping runs from the air handler into the ground and back in one continuous closed loop. Filled with circulating water, the loop uses the earth's constant 72 degree temperature to heat and cool your home. In the winter, the water in the pipes absorbs heat from the earth

and brings it through the system and into your house.

In the summer, the geothermal heat pump reverses itself, pulling heat from your home and placing it in the ground.

Think of the energy efficiency; you're moving heat rather than manufacturing it. Think of the lower maintenance — no outside air conditioning unit exposed to extreme weather conditions. And how about the positive environmental impact — saving energy and protecting the ozone layer by using less refrigerants like freon needed in a traditional system. All this translates to energy cost savings of up to 50 percent on your heating and air conditioning bill.

What's more, you'll save in replacement costs, too. The geothermal heat pump will last twice as long as the conventional air conditioning unit.

You can check out the geothermal heat pump first hand at the Rutenberg Showcase Home during the "Street of Dreams" at Tampa's Avila community, from Oct. 22 through Nov. 27. Admission is $7.

Of course, you won't be able to actually see the geothermal heat pump, but you will be able to "experience" it!

Figure 14-5 Geothermal heating and cooling offers many benefits, although the technology isn't much better known by the public than it was in the early 1990s, when this flyer was produced. *(Copyright 1994 Tampa Electric Company. Use or reproduction of this work is strictly prohibited without prior written consent of Tampa Electric Company.)*

range in our experience). Dollar for dollar invested, geothermal air-conditioning can do more to offset the need for additional power plants and power distribution than any other renewable technology of which we are aware.

Can drilling and trenching wells or loops cause environmental and property concerns? It's true that some of the large-scale drilling projects for geothermal energy (electricity) production have caused small earthquakes and some property damage, but it is important to note that this should not be associated with the relatively shallow and small-scale drilling done for geothermal HVAC systems. As stated previously, at Egg Geothermal, we go to great lengths to leave a customer's property in better shape than we found it, and we don't think the minor excavations we have to do leave any lasting problems.

Is there a lack of education and training for contractors, resulting in poor installations? As we have seen numerous times in this book, there is reason to be concerned here. But that is why it is important to ask for credentials and to make sure proper geothermal engineering is employed in each project. Standards of training are being improved, and there are many excellent firms out there doing fantastic work. We don't feel that a relatively few—ahem—*bad eggs* justify not giving this promising and exciting technology a chance.

Perhaps our greatest challenge is to get the message of geothermal air-conditioning out to the general public. In the early 1990s, I worked with Tampa Electric Company on some marketing materials to spread the word about geothermal.

A facility at the University of South Florida had been outfitted with geothermal HVAC, and we made some print and video ads. The most exciting possibility was the hope of using a celebrity voice. Since the technology takes advantage of the stable temperature in the ground under foot, there was a movement to engage Paul Hogan's character in *Crocodile Dundee*, using some clever line such as, "come to the land down under."

Well, it is almost 20 years later, and we are still waiting for a voice. But the technology works better and more efficiently than ever. So why not come to the land down under?

APPENDIX

Geothermal HVAC Resources

Government

U.S. EPA Energy Star Hotline
1200 Pennsylvania Ave. NW
Washington, DC 20460
888-STAR-YES (888-782-7937)
www.energystar.gov

U.S. Department of Energy's Office of Energy Efficiency and Renewable Energy (EERE)
877-EERE-INF (877-337-3463)
www.eere.energy.gov

Database of State Incentives for Renewable Energy (DSIRE)
Funded by the DOE, DSIRE is a useful source of information on state, local, utility, and federal incentives on renewable energy and energy efficiency.
www.dsireusa.org

Office of Energy Efficiency (Canada)
580 Booth Street
Ottawa, ON K1A 0E4, Canada
613-996-4397
http://oee.nrcan.gc.ca/english

Environment Agency [UK]
Oversees environmental issues in the United Kingdom, including the geothermal industry.
National Customer Contact Centre
PO Box 544
Rotherham S60 1BY, United Kingdom
08708 506 506
enquiries@environment-agency.gov.uk
www.environment-agency.gov.uk

Advocacy and Professional

Air Conditioning Contractors of America (ACCA)
The non-profit, standards-developing organization committed to high-quality, high-efficiency heating and cooling systems in buildings.
2800 Shirlington Road, Suite 300
Arlington, VA 22206
703-575-4477
www.acca.org

Air Conditioning, Heating and Refrigeration Institute (AHRI)
The trade association that represents more than 300 manufacturers of air-conditioning, heating, and commercial refrigeration equipment.
2111 Wilson Blvd, Suite 500
Arlington, VA 22201
703-524-8800
ahri@ahrinet.org
www.ahrinet.org

American Council for an Energy-Efficient Economy (ACEEE)
Nonprofit organization dedicated to advancing energy efficiency.
529 14th Street NW, Suite 600
Washington, DC 20045-1000
202-507-4000
ace3info@aceee.org
www.aceee.org

Earth Energy Society of Canada
Since 1985, the Earth Energy Society of Canada and its predecessor, the Canadian Earth Energy Association, have represented the geothermal heat pump industry in that country.
124 O'Connor, Suite 504
Ottawa, Canada K1P 5M9
613-371-3372
Eggertson@EarthEnergy.ca
www.earthenergy.ca

European Heat Pump Association (EHPA)
Represents the geothermal HVAC industry in Europe.
Renewable Energy House
Rue d'Arlon 63-67
B-1040 Brussels, Belgium
+32 24 00 10 17
info@ehpa.org
www.ehpa.org

Geothermal Exchange Organization (GEO)
A trade association of manufacturers, architects, engineers, heating and cooling businesses, drilling companies, earth loop installers, and other associated businesses; it maintains a directory of industry professionals and active online forums.
1050 Connecticut Avenue NW, Suite 1000
Washington, DC 20036
888-All-4GEO
www.geoexchange.org

Ground Source Heat Pump Association (GSHPA) [UK]
The industry association for the geothermal HVAC industry in the United Kingdom.
National Energy Centre
Davy Avenue, Knowlhill
Milton Keynes, MK5 8NG, United Kingdom
01908 665555
info@gshp.org.uk
www.gshp.org.uk

Heat Pump Centre of the International Energy Agency
An international information service and advocate for heat pumping technologies, applications and markets.
c/o SP Technical Research Institute of Sweden PO Box 857
SE-501 15 BORÅS, Sweden
+46 10 516 5512
hpc@heatpumpcentre.org
http://heatpumpcentre.org

International Ground Source Heat Pump Association (IGSHPA)
Based at Oklahoma State University, IGSHPA is a leading professional association for the geothermal HVAC industry and maintains a list of qualified local installers.
374 Cordell South
Stillwater, OK 74078
405-744-5175
igshpa@okstate.edu
www.igshpa.okstate.edu

U.S. Green Building Council (USGBC)
The community of professionals—organized into regional chapters—working to spread green building by serving as a resource, administering the LEED certification program, and advocating policy.
2101 L Street NW, Suite 500
Washington, DC 20037
www.usgbc.org

Manufacturers

ClimateMaster
A world leader in geothermal HVAC systems.
7300 S.W. 44th Street
Oklahoma City, OK 73179
877-436-6263
www.climatemaster.com

Econar
Manufacturer of GeoSource geothermal heat pumps.
Suite 120—Meridian Business Center
7550 Meridian Circle
Maple Grove, MN 55369
800-4-ECONAR
www.econar.com

NIBE Energy Systems
A leading manufacturer of ground source heat pumps and other HVAC equipment in Northern Europe.
Box 14
Hannabadsvägen 5
28521 Markaryd, Sweden
+46 433-73000
info@nibe.se
www.nibe.eu

Florida Heat Pump
A leading manufacturer of water source and geothermal heat pumps.
601 N.W. 65th Court
Ft. Lauderdale, FL 33309
954-776-5471
heatpump@fhp-mfg.com
www.fhp-mfg.com

WaterFurnace International
Leading manufacturer of geothermal heat pumps.
9000 Conservation Way
Fort Wayne, IN 46809
800-GEO-SAVE
www.waterfurnace.com

Installers

Egg Geothermal
Leading geothermal design, engineering, and contracting services worldwide.
Offices: Tampa, Atlanta
7906 Leo Kidd Ave
Port Richey, FL 34668
866-960-1242
www.egggeothermal.com

Air Perfect, Inc.
Geothermal (and efficient HVAC) installers serving Connecticut and part of New York State.
7 Pearson Avenue
Milford, CT 06460
203-878-0878
http://airperfectinc.com/

Coastal Caisson Corp.
Provides deep foundation services for geothermal installers and others in the construction industry in the southeastern U.S.
13203 Byrd Legg Drive
Odessa, FL 33556
727-536 4748, 800-756-6686
www.coastalcaisson.com
info@coastalcaisson.com

Parrish Services
Geothermal installers serving Virginia and Maryland.
703-330-5748, 800-924-0404
www.parrishservices.com

Search the Geo Exchange directory for a distributor, dealer, or contractor near you:
http://www.geoexchange.org/

Search the IGSHPA directory for an accredited installer near you:
http://www.igshpa.okstate.edu/directory/directory.asp

Index

Note: A page number followed by *f* indicates a figure.

D

E

I

K

L

M

N

O

P

Q

R

S

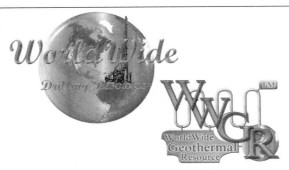

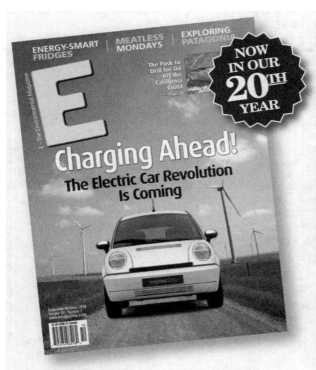